Showtime Kids

National Directory for Parents, Teens, Young Adults, Adults, Talent Schools, Entertainment Industry to:

- Casting Directors
- Casting Agents
- Voice Over Work
- Mailing Labels Index
- Cover Letters and Resumes for Children, Teens, Young Adult Actors/Actresses, and Models
- How to Become a Film/TV Star
- How to Become a Talent Agent

By K. Rucker

ISBN 0-7414-3067-3

Published by:

INFINITY
PUBLISHING.COM
1094 New DeHaven Street, Suite 100
West Conshohocken, PA 19428-2713
Info@buybooksontheweb.com
www.buybooksontheweb.com
Toll-free (877) BUY BOOK
Local Phone (610) 941-9999
Fax (610) 941-9959

Printed in the United States of America

Printed on Recycled Paper

Published April 2006

TABLE OF CONTENTS

A Word from the Author

I have decided to develop this directory because there is a major lack in the showbiz area in employing child actors. Since there has become a great increase in child talent, talent agencies are finding it hard to find work for hundreds of child actors. A gift of talent, not being explored, is waste.

Important

These materials in this book represent the opinions of the author and may not be applicable to all situations. Many circumstances appear similar, but differ in respects, which may be legally significant. In addition, laws and customs change overtime, and by necessity of the lapse in time between the writing and printing of this book, some aspects may be out of date even upon first publication. Accordingly, the author and publisher assume no responsibility for actions taken by readers based upon the advice offered in this book. Each reader should use caution in applying any material contained in this book to his or her specific circumstance, and should seek the advice of an appropriate professional. Author's note: Use your common sense and be careful.

How to Become a Film, Theatre or Television Star

www.kidsworld.com

You see and hear celebrities every day on your television, video, the stage, silver screen, and the radio. Thousands of famous celebrities enjoy the "limelight" of being a performing artist. In fact jobs are quite plentiful in the commercial entertainment industry.

Take a look at the locally produced TV commercials and cable programs that feature adult and child actors/actresses of all ages, shapes and sizes from various ethnic backgrounds. You may not realize that the majority of these local "celebrities" live in your own area, yet you may not even know their names. This directory is design to help give you step by step resources in becoming a working actor/actress.

Training Classes and/or Workshops: Why take training? Acting schools are vocational training institutions that specialize in courses related to theater arts, motion pictures and television arts. The specific curriculum varies rather widely from school to school, but generally you can specialize in the following areas: Acting, Contemporary Theater, Stage Acting, Acting TV, Film, and Radio, Specials performance skills, Improv skills, Voiceover skills and Stunt Acting.

Persistency and determination: In order to be a working actor, you must be willing to handle the auditions. Are your feelings easily hurt or can you stand constant rejection? Now you are ready to become an actor. It is said that if you get hired for one job out of each 10 auditions, you are doing well. That means at least nine rejections in order to secure one paying job. Once you have become a seasoned actor and have built your résumé with a body of work, your opportunities and paycheck will increase. Persistency and determination are key factors as well as choosing a good agent.

Locating an agent: "How do you find an agent? If you know someone who is an actor, either in commercials, film or theater, ask who their agent is. Perhaps one of your co-workers or relatives knows someone who is "in the business." Referrals and word of mouth contacts from working actors are the preferred methods for obtaining a good agent. If you live in a large metropolitan area, look in the yellow pages under the heading of "talent agencies & managers."

"Call each of them and try to set up an appointment. Typically, they may request that you submit a recent photo of yourself along with a cover letter that explains the area of work you in which you are interested, such as commercial, film or TV. Be cordial, polite and, most importantly, persistent! Remember that these agencies may get hundreds of calls each week from potential actors and actresses."

"Make absolutely sure that any agency you contact is a SAG/AFTRA franchised agency. This affiliation with these actors' unions will ensure that your agency can send you on auditions for commercials and/or films, which will air nationally (not just locally, which is considered non-union work). Avoid so-called "talent managers," at least until your career has taken off, because managers typically command as much as 20 percent of your gross income, compared to a talent agent who typically receives 10 to 15 percent of your income from a specific booking."

"Be sure to steer clear of any agency that charges you a separate fee in order to be employed or registered with that agency. Legitimate agencies do not charge their talent up-front fees. Agencies earn commissions based on your earnings from the work performed. In some cases, agents are paid their commissions directly from the producers or sponsors of a particular program, thereby avoiding a commission deduction from your paycheck."

Unions: Joining a union means your access to a lot **more casting calls** and the pay is a lot better. There are three unions for actors and you can find out how you can become a member by going to their Web sites: **Actors' Equity Association**, **American Federation of Television and Radio Artists** and the **Screen Actors Guild**. Remember though, just because you've managed to get into an actor's union, doesn't mean you'll be on the big screen tomorrow.

Headshots: "Once you have secured an agent, the next order of business is obtaining an 8" x 10" (black and white) headshot of yourself from a professional photographer. (Your agent should be able to recommend one.) Expect the photographer to shoot at least two rolls of film (36 pictures per roll). You can expect to get one to two acceptable photos per roll. The final chosen photo should represent you in the best light in the eyes of a casting director.

Consult with your agent for help in choosing the most appropriate headshot based on the type of work you would like to do. Young children's photos typically depict them smiling, although acceptable photos for adults may depict them as either smiling or not. Agents have different preferences regarding types of headshots, so be sure to obtain their opinions before making your final selection."

"You can expect to pay the photographer between $100 to $200 for their services. Reproduction of your master headshot is relatively inexpensive. If you are planning to audition on a weekly basis, you will need to print at least 250 headshots at a cost of between $75 to $200. Your agent will need at least a dozen or more headshots for their files and mailings and you will need to take at least two to three headshots with you to each audition.

Your personal information and your résumé of work performed should be computer-printed on the back of each. Remember that your headshot is all-important — it is your personal calling card, which best represents you and your unique, physical ability to perform the specific role."

Resume: This is a written copy of any **previous** acting work you might have done. It's usually **one page** in length. It lists the roles you've played in Movies, TV or plays, your training (if any) and all the special skills that make up your self. **Special skills** include things like juggling and skateboarding.

Passports: Would you like to network your skills overseas? Get a passport. Also, since some role offers may require filming overseas, passports are needed while filming overseas. It is important to have a passport for acting because it will decrease the risk of being mistaken as of Terrorists.

At a time when state control is diminishing or nonexistent across growing swaths of the world, non-state armed groups pose a little-understood but important challenge to the United States and other nations. This was true before the tragic events of 9/11 and it remains true three years later. In order to meet this premier security challenge of the modern age, we must gain a much deeper understanding of the ways in which criminal states and non-state armed groups interact.

Non-state actors can be roughly divided into four types:

- Insurgents who are engaged in a protracted political and military struggle aimed at weakening or destroying the power and legitimacy of a ruling government;
- Terrorists who spread fear through the threat or use of proscribed violence for political purposes;
- Militias made up of irregular yet recognizable armed forces operating within an ungoverned area or a weak or failing state;
- International criminal organizations engaged in one or more type of criminal enterprise that operates across regions and national borders.

These non-state actors become particularly dangerous when they are hosted by or operate in conjunction with illegitimate states – states that are states only in name. In such cases, the state itself becomes a functioning criminal enterprise while continuing to enjoy – and use for illegal purposes – many of the international benefits of statehood (e.g., the ability to issue recognized diplomatic passports; maintain shipping and airplane registries; control border entry and exit points; and collect taxes).

Auditions: Now you have an agent and pictures. What's next? You simply inform your agent when you will be available for auditions and/or work. Beware. Whether you are an adult actor or a parent of a child actor, you must adjust your regular work schedule, accordingly. Typically, auditions for children take place in the afternoons (after school) and auditions for adults are held in the mornings and afternoons and sometimes on weekends. **Be prepared, cool, calm and relax**. **Be at lease 30 minutes early.**

"The best part of this profession is that you can do it part-time. You decide which days and times of the week you can audition or work. Be realistic with your time. However, if you want a real chance for success in this business, make yourself available for all auditions that come your way. Your agent will love you for it and reward you by keeping you busy. At this point, you may want to enroll in an acting class or seminar on auditioning techniques. It is usually not necessary for children under the age of eight to enroll in acting classes, unless they are very shy or have not had the opportunity of participating in school plays or musicals. (The common complaint by casting directors regarding small children who attend acting classes is that these children appear too rehearsed and not natural during the audition process.)"

The process: "How does the audition process work? A production company, whether for a commercial, film or TV, will contact a specific casting director in your area and notify him/her about the specific character requirements of the production. Perhaps the production company needs three small children between ages five and eight, or adult actors who can portray a married couple, perhaps with children, or several teenagers from various ethnic backgrounds. The casting director or assistant will contact the talent agencies in your area with their request for specific talent that may be able to portray the roles required. The talent agents will review their files and headshots and look for specific talent they feel can best meet the qualifications for the role(s). This is where either you or your children come into the picture."

"Fortunately, an agent can usually send at least three or more of their talent for each specific role being cast. This gives the casting director plenty of talent to review during the audition while increasing your agent's chance for casting one of his/her own talent for the role — hopefully you (or your child). During the audition, you may be requested to perform a simple act without reading one line, or you may be provided a sample or actual script from which to read one or more lines while being taped by a video camera. The tape will be reviewed later by a representative of the production company for final approval of the talent under consideration."

"If the reviewers like what they see, you may receive a "call back. You will be asked, once again, to audition for the job, thus giving the casting director and/or production representative another chance to look you over. Depending on the type of production, you may receive up to two or more callbacks prior to being finally cast. In some cases, you may get the job on the first try!"

Two sons of author Mike Iglesias are professional actors. Brandon, the younger, is co-starring in a new TV series, "Taina," which is scheduled to air on the Nickelodeon channel in January.

Working on Movies as an Extra

www.kidzworld.com

For all of you who are dying to break into the **"industry"** but just haven't been discovered. Instead of waiting around for something to happen, you have to get out there and show the world what you're made of.

How Do You Become an Extra?

The cool part about being an extra is you don't have to have any **formal training** to do it. **Casting Directors** look for people of all ages, depending on what they are working on; **young people** are in high demand. But it's not as easy as just showing up on set and working. All extras need **an agent** who will then tell you about the parts you can go for.

Getting an Agent

Agents are not easy to get. You're going to need to put together **a portfolio** and handing it out to any agent you can think of. Your parents have got to help you out with this because they should like and trust your agent. In your portfolio you'll need **head shots** (which cost between $200-500,) a five-minute **demo tape** (which shows you acting a bit,) and a **resume**. This is where having training helps. While you don't need to have **formal training**, it will help you acquire an agent faster, there's a lot of **competition** out there.
Film school helps or taking **individual courses** at local acting schools if you can't afford (or aren't ready,) to go to film school **full-time**.

A Day in the Life of an Extra

Being an extra is never the same everyday. One day you could be playing **an alien**, the next a **dead creature**, or maybe a kid in high school.

The Extra Factors

Movies have to shoot whenever the time is right - and a lot of the time that means **at night**, outside. Sometimes you have to work from **6 PM to 6 AM**, outside, in the cold. Not fun. And about 70 percent of the time, you'll have to work **over 12 hours** a day! Think you could handle it?

You Might Not Know

Under no circumstances are you allowed to **approach or talk** to celebilities on a set you are working on. So if you think **extra work** is a great way to meet all your Hollywood idols, think again. If a **celebility approaches you**, that's totally cool but you have to remember that this is work and not a movie premiere - so leave those cameras at home!

Making Some Extra Cash

The average extra will make approximately **$10 per hour** for the first eight hours of work. If they need you to stay on set longer, you'll make **more money** as the hours drag on. Every hour after 12 hours of work you'll make approximately **$20 an hour**.

Words of Advice for Future Extras

"Get **an agent**. Even if you only want to work during the summer, get an agent just for the summer. And talk to **your parents** because they're the ones who are going to have to take you to all **the sets** and sit with you while you wait to go on."

ACTOR/MODEL COVER LETTER

Your Name
Your Address
City, State & Zip Code
Phone/Cell/Pager Number
Fax number
Email Address

Date

Agent's Name
Agency Name
Agency Address
City, State & Zip Code

Dear Agent's Name,

I am seeking representation in the commercial, film and/or theater entertainment industries. Enclosed please find my current resume and headshot.

-OR-

I am seeking representation in the modeling industry. Enclosed please find my current composite card and/or tear sheets.

Please contact me if you wish to schedule a personal interview, or if you have any questions regarding my past experience. Thank you for your consideration. I look forward to hearing from you soon.

Sincerely,

Your Full Name

Enclosure(s)

ACTOR COVER LETTER TO CASTING DIRECTOR

Sample Cover Letter

Date

Casting Director's Name/Title
Agency Name
Agency Address
City, State & Zip Code

Dear Casting Director's Name,

I am writing to you as a working actor who would like very much to be considered for work on "Name of the Show." As indicated by the enclosed resume I have experience and/or training in film and stage.

Thank you for your time and consideration. I look forward to having the opportunity to audition for you soon.

Sincerely,

Your Full Name

C/O Agent's Name
Agency Name
Agency Address
City, State & Zip Code

ACTOR/MODEL RESUME

First Name, Last Name
(If represented by an agent) Type Agency Name
Type Agency City and State
Type Agency Phone Number / Fax Number

(If you're not represented by an agent, *type your name, address, email, phone and fax instead)*

(For security purposes it is recommended that you use a cell phone or voice mail number instead of your home number.)

Height:
Eyes:
Weight:
Hair:

TRAINING/WORKSHOPS

Type of Training	Name of Instructor	Name of training facility

SKILLS/SCHOOL ACTIVITIES *(Example Honor Roll Studies)*

(Refer to SAG Skill Sheet to list the skills you are proficient at only.)

(The following categories film, television, commercial/industrial, and theatre; list only the categories that apply to you skills.)

FILM

Name of the films	Principal, Featured, Day Player, Stand-in Extra	Production company or director

TELEVISION

Name of the TV Shows	Principal, Stand-in, Extra	Production company or director

COMMERCIAL/INDUSTRIAL

Name of the Commercial/Industrial	Principal, Featured, Extra	Production Company or director

THEATRE

Name of the Theatre Productions	Lead, Character, Understudy	Production Company or director

PASSPORT UPON REQUEST
WILLING TO TRAVEL WORLDWIDE

SAG Skills Sheet

Accented Actor
Acrobat
Aerialist
Animal Handler (Wild)
Archer
Artist
Arts & Crafts
Badminton Player
Balloonist
Baifaer
Barrel Racer
Bartender
Baseball Player
Basket Weaver & Loomer
Basketball Player
Baton Twirler
Bicyclist (Racing)
Billiard Player
Bird Caller
Blacksmith
Bowler
Boxer
Calligrapher
Carpenter
Ceramic Specialist
Chef
Circus Performer
Clown
Cricket Player
Deep Sea Diver (Hard Hat)
Dealer (Cards, Dice, Roulette)
Dialect Specialist
Diver (Springboard/Platform)

Figure Skater
Fire Eater
Fisherman
Football Player
Glider Pilot (Licensed)
Golfer
Guitarist
Gymnast
Hairdresser
Handball Player
Hang Glider
Helicopter Pilot (Licensed)
Hockey Player
Horse Trainer
Horseback Rider (Multiple Skills)
Hypnotist
Ice Skater
Impersonator (Mimic)
Interpreter
Juggler
Landsailer
Lifeguard (Certified)
Lumberjack
Marksman (Rifles & Pistols)
Martial Arts Specialist (Judo, Karate)
Masseuse • Masseur
Mime
Model
Motorcyclist
Mountain Climber
Painter (House)
Parachutist (Licensed)
Photographer (Still)

Racquetball Player
Rodeo Performer (Multiple Skills)
Roller Skater
Rollerblader
Rope Skipper
Sailor & Boater
Scuba Diver
Sign Language (and Braille)
Silversmith
Skateboarder
Skier (Snow)
Skier (Water)
Skydiver (Licensed)
Snake Handler
Snorkeler
Soccer Player
Squash Player
Stilt Walker
Stone Mason
Stripteaser
Stunt Driver
Surfer
Swimmer
Sword Fighter
Sword Swallower
Tennis Player
Track & Field Athlete
Trampoliner
Triplets
Heavy Equipment Operator
Twins
Unicyclist
Ventriloquist

Dog Sled Racer
Dog Trainer
Driver (Special Licensed)
Dune Buggy Driver
Explosives Handler (Licensed)
Fast Drawer
Female Impersonator
Fencer

Pianist
Pie Thrower
Pilot (Airplane) (Licensed)
Ping Pong Player
Pizza Maker
Polo Player
Pool Player
Puppeteer & Marionetteer

Volleyball Player
Water Ballet Performer
Water Polo Player
Weight Lifter & Body Builder
Welder
Wood Carver
Wrestler
Yodeler

(List only skills that you are proficient at)

YOUNG PERSON'S EMPLOYMENT

Written Consent by Parent or Guardian

For employment as an actor, background performer or extra in the Entertainment Industry

---

Date of Consent (M/D/Y)

Dates of Work:

From:

To:

Information of Child:

Name (Please print in full)

Date of Birth (M/D/Y)

Address (Suite #/ Street/ City)

Postal Code

Phone Number (Please include Area Code)

Alternate Phone Number

Name and Phone Number of Talent Agency (If applicable)

Production Information:

Name

Location, Hours,

Contact Phone Numbers

Parent/Guardian Declaration:

I am the parent of this child . I am the legal guardian of this child .

- I have read and am familiar with the Employment Standards Act & Regulations and Best Practices for the employment of Young People in the Entertainment Industry.
- I agree to supervise the Child or have a Chaperone, nineteen years (19) of age or older, supervise the Child at all times while the Child is on set.
- I am responsible for the Child's well being and safety at all times the Child is working.
- I understand it is my responsibility to ensure that the Child maintains the requirements of her/his educational program.
- I noted the specifics of Location, Hours of work and type of work (actor background performer or extra).

This is my written consent for my child to be employed as an Actor, Background Performer or Extra for a production that adheres to the Employment Standards Act & Regulations and Best Practices for the employment of Young People in the Entertainment Industry.

____

Name (Please print in full)

____

Signature

____

Address (If different than Child)

____

Postal Code

____

Phone Number (If different than Child)

____

Alternate Phone Number

This Written Consent must be obtained by the production and retained as employment records.
(Parent or guardian should retain a copy for own records)

PASSPORTS INFORMATION

Information provided by the U.S. Department of State

How to Apply in Person for a Passport

Should You Apply In Person?

You are applying for a U.S. passport for the first time.	**Yes**
Your previous U.S. passport was lost, stolen, or damaged.	**Yes**
Your previous U.S. passport has expired & was issued more than 15 years ago.	**Yes**
Your previous passport has expired and it was issued when you were under 16.	**Yes**
Your name has changed since your passport was issued and you do not have a legal document formally changing your name.	**Yes**

NOTE:

For All Minors Under Age 14

- Each minor child shall appear in person.
- All applications for children under 14 require both parents' or legal guardians' consent. (See Special Requirements for Children Under Age 14)

For All Minors Age 14 to 17:

- Each minor child shall appear in person.
- For security reasons, parental consent may be requested.
- If your child does not have identification of his/her own, you need to accompany your child and present identification.

To Apply in Person for a U.S. Passport
You MUST:

1. Provide *Application for Passport,* Form DS-11

- To download an application form go to US Passports online.
- Or, forms can be obtained from any passport agency or acceptance facility. (Call to check hours of availability.)

NOTE: Please do **NOT** sign the DS-11 application form until the Passport Acceptance Agent instructs you to do so.

2. Present Proof of U.S. Citizenship

You may prove U.S. Citizenship with any one of the following:

Previous U.S. Passport (mutilated, altered, or damaged passports are not acceptable as evidence of U.S.

citizenship.)

Certified birth certificate issued by the city, county or state

NOTE: A certified birth certificate has a registrar's raised, embossed, impressed or multicolored seal, registrar's signature, and the date the certificate was filed with the registrar's office, which must be within 1 year of your birth

- Consular Report of Birth Abroad or Certification of Birth
- Naturalization Certificate
- Certificate of Citizenship

A Delayed Birth Certificate filed more than one year after your birth may be acceptable if it:

- Listed the documentation used to create it **and**
- Signed by the attending physician or midwife, or, lists an affidavit signed by the parents, or shows early public records.

If you do NOT have a previous U.S. passport or a certified birth certificate, you will need:

1. **Letter of No Record**

   Issued by the State with your name, date of birth, which years were searched for a birth record and that there is no birth certificate on file for you.

2. **AND as many of the following as possible:**

   - Baptismal certificate
   - Hospital birth certificate
   - Census record
   - Early school record
   - Family bible record
   - Doctor's record of post-natal care

NOTES: These documents **must** be early public records showing the date and place of birth, preferably created within the first five years of your life.

You may also submit an Affidavit of Birth, form DS-10, from an older blood relative, i.e., a parent, aunt, uncle, sibling, who has personal knowledge of your birth. It must be notarized or have the seal and signature of the acceptance agent.

If you were born abroad AND do not have a Consular Report of Birth Abroad or Certificate of Birth on file, you will need:

1. If you claim citizenship through birth abroad to one U.S. citizen parent:

   - Foreign birth certificate,
   - Proof of citizenship of your U.S. citizen parent, AND
   - An affidavit of your U.S. citizen parent showing all periods and places of residence or physical presence in the United States and abroad before your birth.

2. If you claim citizenship through birth abroad to two U.S. citizen parents:

   - Your foreign birth certificate,
   - Parent's marriage certificate, AND
   - Proof of citizenship of your U.S. parents and an affidavit of your U.S. citizen parents showing all periods and places of residence of physical presence in the United States and abroad before your

birth.

Click here for additional information on documentation of U.S. citizens born abroad who acquire citizenship at birth

NOTE: The following are NOT proof of citizenship

- Voter registration cards
- Army discharge papers

We have new information on foreign-born children adopted by U.S. citizens.

NOTE: If you travel extensively, you may request a larger, 48-page passport at no additional cost. To do so, please attach a signed request for a 48-page passport to your application.

FOR MINORS UNDER THE AGE OF 14:

The citizenship evidence submitted for minors under the age of 14 must list both parents' names. Read more information on the citizenship requirements for minors under the age of 14.

3. Present Proof of Identity

You may prove your identity with any one of these, if you are recognizable:

- Previous U.S. passport (mutilated, altered, or damaged passports are not acceptable as proof of identity.)
- Naturalization Certificate
- Current, valid
  - Driver's license
  - Government ID: city, state or federal
  - Military ID: military and dependents

NOTE: Your Social Security Card does NOT prove your identity.

If none of these are available , you will need:

1. Some signature documents, not acceptable alone as ID

   (ex: a combination of documents, such as your Social Security card, credit card, bank card, library card, etc.)

   AND

2. A person who can vouch for you. He/she must
   - Have known you for at least 2 years,
   - Be a U.S. citizen or permanent resident,
   - Have valid ID, and
   - Fill out a Form DS-71 in the presence of a passport agent.

FOR MINORS UNDER THE AGE OF 14:

Each minor child shall appear in person. Both parents or legal guardians must present evidence of identity when they apply for a minor under the age of 14. Read more information on the identity requirements for minors under the age of 14.

FOR MINORS 14 to 17:

- Your child MUST appear in person.

- For security reasons, parental consent may be requested.
- If your child does not have identification of his/her own, you need to accompany your child, present identification and co-sign the application.

4. Provide Two Passport Photos

Your photographs must be:

- 2x2 inches in size
- Identical
- Taken within the past 6 months, showing current appearance
- Color
- Full face, front view with a plain white or off-white background
- Between 1 inch and 1 3/8 inches from the bottom of the chin to the top of the head
- Taken in normal street attire
    - Uniforms should not be worn in photographs except religious attire that is worn daily.
    - Do **not** wear a hat or headgear that obscures the hair or hairline.
    - If you normally wear prescription glasses, a hearing device, wig or similar articles, they should be worn for your picture.
    - Dark glasses or nonprescription glasses with tinted lenses are not acceptable unless you need them for medical reasons. A medical certificate may be required.

Click here for information on acceptable digitized photos.

Click here for detailed information for professional photographers.

NOTE: Vending machine photos are not generally acceptable

5. Pay the Applicable Fee

Our 13 Passport Agencies Accept:

For Passport Application and Execution Fee:

- Credit Cards – VISA, MasterCard, American Express, Discover
- Debit/Check cards (not ATM cards)
- Checks, money orders, bank drafts

Our over 7,000 Passport Acceptance Facilities Accept:

For Passport Application Fee:

- Personal checks, money orders, and bank drafts at all locations
- Exact cash at some locations

For Execution Fee:

- Money orders and bank drafts at all locations

- Personal checks and exact cash at some locations
- Credit cards at U.S. Postal Facilities and some other locations

NOTE: When applying at one of the thousands of designated application acceptance facilities in the U.S., you pay the **application processing fee** to the "U.S. Department of State" and the **execution fee** to the facility where you are applying. When applying at a Regional Passport Agency, both fees are combined into one payment to the "U.S. Department of State."

Expedite Fee: (See How to Get Your Passport in a Hurry.)

6. Provide a Social Security Number

If you do not provide your Social Security Number, the Internal Revenue Service may impose a $500 penalty. If you have any questions please call your nearest IRS office.

Where to Apply for a Passport in the U.S.:

- **BOSTON** Passport Agency
  Thomas P. O'Neill Federal Building
  10 Causeway Street,
  Suite 247
  Boston, MA 02222-1094

Hours: 8:30 a.m. - 4:30 p.m., local time, M-F, excluding Federal holidays
Automated Appointment Number: 1-877-487-2778

NOTE: Boston Passport Agency serves customers who are traveling within 2 weeks (14 days), or who need foreign visas for travel. An appointment is required.

- **CHICAGO** Passport Agency
  Kluczynski Federal Building
  230 S. Dearborn Street,
  18th Floor
  Chicago, IL 60604-1564

Hours: 9:00 a.m. - 4:00 p.m., local time, M-F, excluding Federal holidays
Automated Appointment Number: 1-877-487-2778

NOTE: Chicago Passport Agency serves customers who are traveling within 2weeks (14 days), or who need foreign visas for travel. An appointment is required.

- **COLORADO** Passport Agency
  Cherry Creek III
  3151 South Vaughn Way, Suite 600
  Aurora, CO 80014

Hours: 8:30 a.m. - 3:30 p.m., local time, M-F, excluding Federal holidays
Automated Appointment Number: 1-877-487-2778

NOTE: Colorado Passport Agency serves customers who are traveling within 2weeks (14 days), or who need foreign visas for travel. An appointment is required.

- **CONNECTICUT** Passport Agency
  50 Washington Street
  Norwalk, CT 06854

Hours: 9:00 a.m. - 4:00 p.m., local time, M-F, excluding Federal holidays
Automated Appointment Number: 1 (877) 487-2778

NOTE: Connecticut Passport Agency serves customers who are traveling within 2 weeks (14 days), or who need foreign visas for travel. An appointment is required.

- **HONOLULU** Passport Agency
  Prince Kuhio Federal Building
  300 Ala Moana Blvd.
  Suite 1-330
  Honolulu, HI 96850

Hours: 8:30 a.m. - 3:30 p.m., local time, M-F, excluding Federal holidays
Automated Appointment Number: 1 (877) 487-2778

NOTE: Honolulu Passport Agency serves customers who are traveling within 2 weeks (14 days), or who need foreign visas for travel. An appointment is required.

- **HOUSTON** Passport Agency
  Mickey Leland Federal Building
  1919 Smith Street
  Suite 1400
  Houston, TX 77002-8049

Hours: 8:30 a.m. - 3:30 p.m., local time, M-F, excluding Federal holidays
Automated Appointment Number: 1-877-487-2778

NOTE: Houston Passport Agency serves customers who are traveling within 2 weeks (14 days), or who need foreign visas for travel. An appointment is required.

- **LOS ANGELES** Passport Agency
  Federal Building
  11000 Wilshire Blvd.
  Suite 1000
  Los Angeles, CA 90024-3615

Hours: 8:00 a.m. - 3:00 p.m., local time, M-F, excluding Federal holidays
Automated Appointment Number: 1-877-487-2778

NOTE: Los Angeles Passport Agency serves customers who are traveling within 2 weeks (14 days), or who need foreign visas for travel. An appointment is required.

- **MIAMI** Passport Agency
  Claude Pepper Federal Office Building
  51 SW First Avenue
  3rd Floor
  Miami, FL 33130-1680

Hours: 8:00 a.m. - 3:00 p.m., local time, M-F, excluding Federal holidays
Automated Appointment Number: 1-877-487-2778

NOTE: Miami Passport Agency serves customers who are traveling within 2 weeks (14 days), or who need foreign visas for travel. An appointment is required.

- **NEW ORLEANS** Passport Agency
  One Canal Place (corner of Canal and North Peters Streets)
  365 Canal Street, Suite 1300
  New Orleans, LA 70130-6508

Hours: 8:30 a.m. - 3:30 p.m., local time, M-F, excluding Federal holidays
Automated Appointment Number: 1-877-487-2778

NOTE: If you are **not traveling in the next 6 weeks**, there is no action required from you at this time. Please visit this website periodically for further updates.

NOTE: New Orleans Passport Agency serves customers who are traveling within 2 weeks (14 days), or who need foreign visas for travel. An appointment is required.

- **NEW YORK** Passport Agency
  376 Hudson Street
  New York, NY 10014

Hours: 7:30 a.m. - 3:00 p.m., local time, M-F, excluding Federal holidays
Automated Appointment Number: 1-877-487-2778

NOTE: New York Passport Agency serves customers who are traveling within 2 weeks (14 days), or who need foreign visas for travel. An appointment is required.

- **PHILADELPHIA** Passport Agency
  U.S. Custom House
  200 Chestnut Street
  Room 103
  Philadelphia, PA 19106-2970

Hours: 9:00 a.m. - 4:00 p.m., local time, M-F, excluding Federal holidays
Automated Appointment Number: 1-877-487-2778

NOTE: Philadelphia Passport Agency serves customers who are traveling within 2 weeks (14 days), or who need foreign visas for travel. An appointment is required.

- **SAN FRANCISCO** Passport Agency
  95 Hawthorne Street
  5th Floor
  San Francisco, CA 94105-3901

Hours: 8:00 a.m. - 3:00 p.m., local time, M-F, excluding Federal holidays
Automated Appointment Number: 1-877-487-2778

NOTE: San Francisco Passport Agency serves customers who are traveling within 2 weeks (14 days), or who need foreign visas for travel. An appointment is required.

- **SEATTLE** Passport Agency
  Henry Jackson Federal Building
  915 Second Avenue
  Suite 992
  Seattle, WA 98174-1091

Hours: 8:00 a.m. - 3:00 p.m., local time, M-F, excluding Federal holidays
Automated Appointment Number: 1-877-487-2778

NOTE: Seattle Passport Agency serves customers who are traveling within 2 weeks (14 days), or who need foreign visas for travel. An appointment is required.

- **WASHINGTON** Passport Agency
  1111 19th Street, N.W.
  First Floor, Sidewalk Level
  Washington, D.C. 20036

Hours: 8:00 a.m. - 3:00 p.m., local time, M-F, excluding Federal holidays
Automated Appointment Number: 1 (877) 487-2778

NOTE: Washington Passport Agency serves customers who are traveling within 2 weeks (14 days), or who need foreign visas for travel. An appointment is required.

- **SPECIAL ISSUANCE** Agency
  1111 19th Street, N.W. Suite 200
  Washington, D.C. 20036

Hours: 8:15 a.m. - 4:30 p.m., local time, M-F, excluding Federal holidays

NOTE: Applications for Diplomatic, Official, and No-Fee passports

To serve you better, we have 2 Passport Processing Centers designed to help us meet our increasing demand for passports.

Please note that these Centers are NOT open to the public.

- **CHARLESTON** Passport Center
  1269 Holland Street
  Charleston, SC 29405

Hours: 8:00 a.m. - 4:00 p.m., local time, M-F, excluding Federal holidays

- **NATIONAL** Passport Center
  31 Rochester Avenue
  Portsmouth, NH 03801-2900

Hours: 9:00 a.m. - 4:00 p.m., Monday through Friday, excluding Federal holidays

U.S. Postal Office in your area may do passports. Check with the Main Headquarters for locations.

Acting Terms You Need To Know

You must walk into an audition with a **basic knowledge** of show business.
These are the definitions for need-to-know acting terms!

A.D. - The Assistant Director.

AD LIB - Extemporaneous delivery without relying on a prepared script.

ADR - Automated Dialogue Replacement. Dialogue added to a scene in post production. Also called "looping".

AEA - Actors' Equity Association; also called "Equity". SAG's sister union which represents stage actors. See also BAE, CAE and MEAA.

AFI - The American Film Institute

A.C.C.T. - Association of Canadian Craftspeople

ACTRA - Alliance of Canadian Cinema Television and Radio Artist

AFTRA - American Federation of Television and Radio Artists. Represents radio artists and news broadcasters, and, in earlier times, television performers. In more recent times, however, television performers may be represented by either AFTRA or SAG, depending on the producer's contract. Discussions about merging the two organizations have been ongoing for several decades; recent Television & Film and Commercial Contracts have been jointly negotiated.

AGMA - American Guild of Musical Artists

AGVA - American Guild of Variety Artists

AMPTP - Alliance of Motion Picture and Television Producers

ART DIRECTOR - Person who conceives and designs the sets.

ATA - Association of Talent Agents

ACTION! - The command from the director for the scene to begin. It also means that the camera is rolling.

ACTOR - This would be you. If you get the part, you'll be able to play other people - and get **paid**. It's one of the best jobs in the world.

AGENT - This is your representation, the person who acts as your contact with the **film company**, and they will most likely have gotten you your audition. They'll also negotiate how much **money** you'll make if you do get a part, and **never, ever forget** you and your agent work for each other.

AUDITION - A tryout for a film, TV or stage role. Usually auditions involving reading from the script, but can also require improvisation.

AVAIL - A courtesy situation extended by performer or agent to a producer indicating availability to work a certain job. Avails have no legal or contractual status.

BACKGROUND - The Extra performers. On the set, "Background!" is a verbal cue for the Extras to start their action.

BACK TO ONE! - The verbal cue for performers to return to the mark where they started the scene.

BAE - British Actors Equity

BEAUTY SHOT - On TV soaps, the shot over which the credits are rolled.

BEST BOY - The assistant to the Chief Electrician, or Head Gaffer.

BILLING - The order of the names in the title of opening credits of a film or TV show.

BIO - Short for "biography". A resume in narrative form, usually for a printed program or press release.

BLOCKING - The actual physical movements by performers in any scene. Also can refer to the movements of the camera.

BOOKING - A firm commitment to a performer to do a specific job.

BOOM - An overhead microphone, usually on an extended pole. The Boom Operator is the member of the sound department responsible for holding the boom pole, with mic attached, over and sometimes under the actors. Also usually responsible for placing radio mics on actors.

BLUE SCREEN - Shooting in a studio against a large blue or green backdrop, which allows a background to be superimposed later on the final image. The actors must imagine the set they are on and be aware of the limitations of their movements. Casting Workbook's Audition studio in their Vancouver location is a Blue Screen.

BREAKAWAY - Specially designed prop or set piece that looks solid but shatters easily.

BREAKDOWN - A summary description of a script prepared by or for the casting director often including the names of the director, producer, network or studio, together with audition location and times, storyline and roles available for casting in a production. These are, and have traditionally been, provided only to qualified talent agents. Breakdowns are posted on the Casting Workbook by the Casting Director and go out to as many as 1000 agents in 20 cities. See also Casting Notices.

BRIEF - The Australian equivalent to the Breakdown.

BROWSER - See Web Browser

CAEA - Canadian Actors Equity Association

CACHE - Your Web browser's cache, which contains the most recent Web files that you have downloaded and which is physically located on your hard disk.

CALLBACK - This is what happens after your first audition, if they decide that they'd like to see you again. You'll be **invited** to audition for the same part again, but this time the **Director and Producer** will most likely be present.

CALL SHEET - A sheet containing the cast and crew call times for a specific day's shooting. Scene numbers, the expected day's total pages, locations, and production needs are also included.

CALL TIME - The actual time an actor is due on the set.

CAMERA CREW - With the D.P. (Director of Photography) as its chief, this team consists of the camera operator, the first assistant camera operator (focus puller), the second assistant camera operator (film loader and clap stick clapper) and the dolly grip.

CAMERA OPERATOR - The member of the camera crew who actually looks through the lens during a take. Responsible for panning, tilting and keeping the action within the frame.

CASTING AGENT - This is the person that talks to your agent about who in your agent's roster might be good for the part. (This, you hope, will always be you.) They work directly with the director and producer to fill in all the roles available on any given project.

CASTING CALL - This is when the **casting agent** gets a notice saying that a movie is coming to town, and that they'll be looking at **local talent**. Then the agent calls you, (if your agent thinks the part might be right for you.) Then, you get your sides and your audition time.

CASTING DIRECTOR - The producer's representative responsible for choosing performers for consideration by the producer or director.

CASTING FACILITY - A studio or space used by one or more casting directors for holding audition taping sessions. Many casting directors have their own casting facility and others rent facilities for their auditions as required.

CASTING NOTICE - Similar in format to a Breakdown, the casting notice is not restricted to agents only. They are distributed to actors, agents and the public, much the same as a posting in a newspaper.

CATTLE CALL - The actual time you are due on the set

CATERER - Responsible for breakfast, lunch and dinner on a set. Different from Craft Services.

CD-ROM - A compact disk that holds text, music and images. One of the principle new venues for interactive video games as well as for full motion video films. Acting for CD-ROM's is a new arena for actors.

CDS - The Casting Delivery Service run by Casting Workbook in Toronto as a daily 5 point drop-off courier and package delivery augmenting the electronic service.

CFTPA - Canadian Film & Television Production Association. A non-profit, trade association representing almost 400 Canadian production companies involved in television, film and interactive media.

CHANGES - Outfits worn while performing.

CHEAT - The actor's adjustment of body position away from what might be absolutely "natural" in order to accommodate the camera; can also mean looking in a different place from where the other actor actually is.

CHECKING THE GATE! - A verbal command to check the lens on the camera; if the lens is OK the cast & crew will move on to the next scene or shot.

CHIEF ELECTRICIAN - Heads the electrician crew; also called the Gaffer.

CINEMATOGRAPHER - Director of Photography

CLOSE-UP (CU) - Camera term for tight shot of shoulders and face.

COLD READING - Unrehearsed reading of a scene, usually at an audition.

COMMISSION - Percentage of a performer's earnings paid to agents or managers for services rendered.

COMPOSITE - A series of photos on one sheet representing an actor's different looks.

CONFLICT - Status of being paid for services in a commercial for one advertiser, thereby contractually preventing performing services in a commercial for a competitor.

COOKIE - A cookie is information that a Web site puts on your hard disk so that it can remember something about you at a later time.

COPY - The script for a commercial or voice over.

COVERAGE - All camera shots other than the master shot; coverage might include two-shots and close-ups.

CRAFT SERVICES - On-set beverage and snack table. Different from the Caterer

CRANE SHOT - A camera shot raised over or above the set or the action.

CRAWL - Usually the end credits in a film or TV shot which "crawl" up the screen.

CREDITS - Opening names in a film or TV show; also refers to a one's performance experience listed on a resume or in a program.

CSA - Casting Society of America. Professional society of Theatrical (Film, TV, Stage) Casting Directors.

CUE - Hand signal by the Stage Manager

CUT! - The verbal cue for the action of the scene to stop. At no time, may an actor call, "cut!"

CUTAWAY - A short scene between two shots of the same person, showing something other than that person.

DAILIES - Screening of footage before it is edited.

DAY PLAYER (DAY PERFORMER) - A principal performer hired on a daily basis, rather than on a longer - term contract.

DAYTIME DRAMA - Soap opera.

DIRECTOR - This is the person who is in charge of what the <u>actors</u> do, who shapes the story that you end up seeing on the big screen. If a movie had a parent, the director would be the **dad**.

DEMO TAPE - An actor's audio or video tape that agents use for audition purposes. These are now going digital and are being uploaded to the Casting Workbook saving duplication and shipping costs for agents and their actors.

DEMO REEL - A **video tape** of all the previous film work you have done. It usually contains different types of stuff, like <u>happy</u> scenes and <u>sad</u> scenes and <u>funny</u> scenes. They are usually about **five minutes** in length and should show as much of the different sides to your talent as possible.

DGA - Directors Guild of America.

DGC - Directors Guild of Canada

DIALECT - A distinctly regional or linguistic speech pattern.

DIALOGUE - The scripted words exchanged by performers.

DIGITIZING - The process of converting something into a form for use in a computer

DIRECTOR - The coordinator of all artistic and technical aspects of any production.

DIRECTOR OF PHOTOGRAPHY (D.P. or D.O.P) - Supervises all decisions regarding lighting, camera lenses, color and filters, camera angle set-ups, camera crew and film processing.

DOLLY - A piece of equipment that the camera sits on to allow mobility of the camera.

DOLLY GRIP - The crew member who moves the dolly.

DOUBLE - A performer who appears in place of another performer, i.e., as in a stunt.

D.P. - Director of Photography or Cinematographer.

DRESS THE SET - Add such items to the set as curtains, furniture, props, etc.

DRIVE-ON PASS - In Los Angeles, a pass to drive onto and park on a studio lot.

DUPE - A duplicate copy of a film or tape; also, a "dub"

8x10 - Commonly used size of a performer's photos, usually in black and white.

18-TO-PLAY-YOUNGER - A performer legally 18 years old, who can convincingly be cast as a younger age.

E. I. C. - Entertainment Industry Coalition

ELECTRICIAN - In film, crew members who place lighting instruments, focus, gel and maneuver the lights.

EMPLOYER OF RECORD (EOR) - The company responsible for employment taxes, unemployment benefits and workers compensation coverage.

ENCODING - Converting a digital file, usually audio or video, into a specific format. Ie: MP3, AVI, MOV etc.

EQUITY - Actors Equity Association (see AEA, BAE , CAEA and MEAA) Union representing stage actors.

EQUITY WAIVER - In Los Angeles, 99-seat (or less) theatres which were otherwise professional, over which Equity waived contract provisions under certain circumstances. Now officially called "Showcase code", the term "Equity waiver" is still used informally.

EXECUTIVE PRODUCER - Person responsible for funding the production.

EXT. (Exterior) - A scene shot outside.

EXTRA - Background performer, used only in non-principal roles. A non-speaking part, or window dressing, as some people call it. Being an extra is a great way to start making money and learn about the process of **movie making**. And, you never know, you might just get chosen to do more. "Who's ready for their close-up?"

FEATURE - Now this is where you want to be. If you want to be famous, want to make lots of dough, well, the feature is where it's at. It's the full-length, big money Hollywood-style movie. And, if you're lucky, you might get to have **lunch** with some stars.

FICA - Social Security taxes (Federal Insurance Corporation of America).

FIELD REP. - Union staff member who ensures contractual compliance on sets.

FIRST A.D. - First Assistant Director; person responsible for the running of the set. Gives instructions to crew and talent, including calling for "first team," "quiet," "rehearsal," and "take five."

FIRST ASS'T. CAMERA OP. - First Assistant Camera Operator is responsible for focusing the camera lens during the shooting of a scene; also known as the Focus Puller.

FIRST TEAM - The production term for the principal actors in a scene.

4-A's - Associated Actors and Artistes of America; umbrella organization for SAG, AFTRA, Equity and other performers' Unions.

FORCED CALL - A call to work less than 12 hours after dismissal on the previous day. See TURNAROUND.

FOREGROUND CROSS - Action in a scene in which an Extra Performer passes between the camera and the principal actors; sometimes called a "wipe".

FX (Effects) - Special Effects.

GAFFER - The Chief Electrician.

GOLDEN TIME - Contractually called 16 Hour Rule Violation for Extra Performers, is overtime, after the 16th hour, paid in units of one full day per hour.

GRIPS - Members of the film crew who are responsible for moving set pieces, lighting equipment, dolly track and other physical movement of equipment.

HEADSHOT - This is an 8x10 picture of your face. Remember to keep your picture **current** and make sure your photo looks like you. If you change looks, be sure to change photos.

HAND MODEL - A performer whose hands are used to double for others.

HIATUS - Time during which a TV series is not in production

HOLDING - The designated area to which the Extra Performers report and stay while waiting to go on set.

HONEY WAGON - A towed vehicle containing one or more dressing rooms, as well as crew bathrooms.

IATSE - International Alliance of Theatrical Stage Employees; the union which represents most off - camera crew members.

INDUSTRIAL - Non-broadcast film or video, usually of an educational nature

INSERTS - Shots, usually close -ups of hands or close business, inserted into previously shot footage.

INT. (Interior) - A scene shot indoors.

"IN" TIME - The actual call time or start time; also, return time from a break.

LINE PRODUCER - The producer responsible for keeping the director on time and budget; generally the most visible producer actually on the set.

LONG SHOT (LS) - A camera shot which captures the performer's full body.

LOOPING - An in-studio technique used to fix dialogue already performed during principal photography by matching voice to picture.

MARK - The exact position(s) given to an actor on a set to insure that he/she is in the proper light and camera angle; generally marked on the ground with tape or chalk.

MARKER! - A verbal cue that the take has been identified on camera both verbally and with the slate board.

MASTER SHOT - A camera shot that includes the principal actors and relevant background activity; generally used as a reference shot to record the scene from beginning to end before shooting close-ups, over-the-shoulders, etc.

MATCHING ACTIONS - The requirement that the actor match the same physical movements in a scene from take to take in order to preserve the visual continuity.

MEAA - Media Entertainment and Arts Alliance (Australia)

MEAL PENALTY - A fee paid by the producer for the failure to provide meals or meal breaks as specified by the contract.

MIXER - Chief of the sound crew; responsible for the quality of the sound recording on a shoot.

MODEM - A device that converts digital signals from a computer or other digital device, into a form that can be transmitted across copper wires. It also re-converts them when received from another device.

MOS (Mit Out Sound/Motion Only Shot) - Any shot without dialogue or sound recording. Also sometimes called S.O.C. , silent on-camera.

M.O.W. - Movie of the week

N.A.B.E.T - National Association of Broadcasting Employees & Technicians

ND MEAL (NON DEDUCTABLE MEAL) - A 15 minute meal break provided to actors by the production company to bring actors in sync with crew break time. It must be completed within 2 hours of performers call time.

NETWORK - In the most generic sense, a group of computers connected to each other allowing communication and a sharing of information. Also may refer to a Television Network such as Fox, ABC etc...

NIGHT PREMIUM - A surcharge for certain work performed after 8 p.m.

OFF-CAMERA (OC or OS) - Dialogue delivered without being on screen.

OPEN CALL - This is just like a **casting call**, but anybody - whether they have an agent or not - can come in off the street. These are often used to cast big extra scenes, but you never know, you could get **discovered**!

OPERATING SYSTEM - The group of programs that allows users to interact with a computer. Primarily Microsoft Windows, Mac OS or Linux.

OUT OF FRAME - An actor outside the camera range.

"OUT" TIME - The actual time when you are released after you have changed out of wardrobe and make- up.

OVER-THE-SHOULDER - A shot over the shoulder of one actor, focusing entirely on the face and upper torso of the other actor in a scene; generally shot in pairs so both actors expressions can later be edited together.

OVERDUBBING - In studio singing or voice work, the process of laying one soundtrack over another.

OVERTIME (OT) - Work extending beyond the contractual work day.

P.A. - Production Assistant.

PAN - A camera shot which sweeps from side-to-side.

PAYMASTER - An independent talent payment service acting as the employer of record.

PENSION & HEALTH PAYMENT - An additional amount of money paid by the employer to cover employee benefits under union contract

PER DIEM - Fee paid by producer on location shoots to compensate performer for expenditures for meals not provided by the producer.

PERIPHERAL - Any device that is optionally attached to a computer. Mouse, keyboard, scanner, printer etc.

PHOTO DOUBLE - An actor cast to perform on camera in place of another.

PICK UP - Starting a scene from a place other than the beginning.

PICTURE'S UP! - Warning that the sequence of cues to shoot a scene is about to begin.

PILOT - This is the term used for the **first episode** of a television series. Basically, there is an idea for a TV show, the screenwriter writes, the actors act and the episode of the show gets made. If the **test audience** likes it, the pilot gets picked up by a studio (Fox, NBC, ABC, UPN - whatever,) that will make at least **one season** or about 13 episodes of the series. If it isn't well liked, it won't get picked up and the pilot itself may never see the light of day. The cool thing is that you get paid either way - NICE!

PLATFORM - See OPERATING SYSTEM.

POV SHOT - Point-of-View shot; camera angle from the perspective of one actor.

POST-PRODUCTION - The phase of filmmaking that begins after the film has been shot. Includes scoring, sound and picture editing, titling, dubbing, and releasing.

PRE-PRODUCTION - The phase of filmmaking before shooting begins; includes writing, scouting locations, budgeting, casting, hiring crews, ordering equipment and creating a shooting schedule.

PRINCIPAL - A performer with lines.

"PRINT!" - A call from the director at the end of a take that that particular take is good enough be printed.

PRODUCER - Often called the Line Producer; this person (or people) acquires the cash, builds the crew and gets people together. The producer is a **problem solver** that gets things done. The producer is often the money behind the project, so it's really good if they like you. The producer would be the **mom**.

PRODUCTION COMPANY - The company actually making the film or television show.

PROPS - Any objects used by actors in a scene.

PSA - Public Service Announcement.

RESIDUAL - The fee paid to a performers for rebroadcast of a commercial, film or TV program

RESUME - List of credits, usually attached to an 8x10 or composite.

REWRITE - Changes in the script, often using color-coded pages to indicate most current version.

RIGHT-TO-WORK-STATES - Those states which do not honor certain union provisions.

ROLLING! - The verbal cue for the camera film and audio tape to start rolling.

ROOM TONE - A sound recording (sometimes made upon completion of a scene) to record existing noise at the location. Also called "wild track".

SAG - Screen Actors Guild.

SCALE - Minimum payment for services under union contracts.

SCREENWRITER - This is the person who wrote the story or script for the movie. Did you know that sometimes there are **multiple screenwriters** for one movie? For example, **X2** had three screenwriters.

SCENES - The Australian equivalent to Sides. See Sides below

SCRIPT - The written form of a screenplay, teleplay, radio or stage play.

SCRIPT SUPERVISOR - The crew member assigned to record all changes or actions as the production proceeds.

SDI - State Disability Insurance.

SECOND ASSISTANT DIRECTOR - Often two or three on a set, they handle checking in the talent, insuring proper paperwork is filed, distribute script revisions. Actors check in with the 2nd A.D. upon arrival on the set.

SECOND TEAM! - The verbal cue for the stand - ins to come to the set and be ready to stand in.

SEGUE - In film or tape editing, a transition from one shot to another.

SET - The immediate location where the scene is being filmed.

SET-UP - Each time the camera changes position.

SFX - Sound effects.

SIDES - Pages or scenes from a script, used in auditions or (if on a film set) those scenes being shot that day. In Australia, Sides are called Scenes. These are the lines from the **script** that you'll be using for your audition. They can be anywhere from **four** lines to **fourteen** pages, maybe more, depending on the part you're up for. You'll get your sides from your **agent** before your audition. Your agent gets them from the casting agent, who gets them from the **production company**.

SIGNATORY - An employer who has agreed to produce under the terms of a union contract.

SLATE - A small chalkboard and clapper device, often electronic, used to mark and identify shots on film for editing; also the process of verbal identification by a performer in a taped audition (e.g., "Slate your name!").

SPEED! - A verbal cue that the audio tape is up to speed for recording.

SPIKING THE LENS - Looking directing into the lens during a scene; since it destroys the illusion of realism, actors should never spike the lens unless specifically directed to do so for specific effect.

STAGE RIGHT - To the performer's right side, to the audience's left side. Likewise, STAGE LEFT is to the performer's left, the audience's right. Stage directions are for actors, not audiences, therefore they are always given from the actor's point of view to the audience.

STANDARD UNION CONTRACT - The standard format/contract approved by the Unions and offered to performers prior to the job.

STANDARDS & PRACTICES - The network TV censorship departments.

STAND-INS - Extra Performers used as substitutes for featured players, for the purpose of setting lights and rehearsing camera moves; also known as the second team.

"STICKS" - Slate or clapboard.

STUDIO - A building, recording room or sound stage which accommodates film or TV production.

STUNT COORDINATOR - The person in charge of designing and supervising the performance of stunts and hazardous activities.

STUNT DOUBLE - A stunt person who performs stunts for a principal.

STUNTPERSON - A specially trained performer who actually performs stunts.

SUBMISSION - An agent's suggestion to a casting director for a role in a certain production.

SW - A notation on a call sheet that an actor is starting on that day and working on that day.

SWF - A notation on a call sheet that an actor is starting, working, and finished on that day.

SWEETENING - In singing/recording, the process of adding additional voices to previously recorded work.

SYNDICATION - Selling TV programs to individual stations rather than to networks.

TAFT-HARTLEY - A federal statute which allows 30 days after first employment before being required to join a Union.

TAKE - The clapboard indication of a shot "taken" or printed.

"TAKE 5" - The announcement of periodic five minute breaks.

T.A.M.A.C. - Talent Agents and Managers' Association of Canada

TELEPROMPTER - The brand name of a device which enables a broadcaster to read a script while looking into the camera lens.

THEATRICAL - TV shows or feature film work, as opposed to commercials.

THREE BELLS! - An audible warning for QUIET because a scene is about to be filmed.

TIGHT SHOT (Go in Tight) - Framing of a shot with little or no space around the central figure(s) of feature(s); usually a close-up.

TILT - The up and down movement of a camera.

TIME-AND-A-HALF - Overtime payment of 1 1/2 times the hourly rate.

TMA - Talent Managers Association (Los Angeles based)

TRACKING SHOT - A shot taken while the camera is moving, either on a dolly or a mounted on a moving vehicle.

TRADES - Short for "trade papers" - The newspapers and periodicals such as the Hollywood Reporter and Variety that specifically feature information on the entertainment industry.

TURNAROUND - (a) The number of hours between dismissal one day and call time the next day. (b) To shoot a scene from another direction.

TWO-SHOT - A camera framing two persons.

UNDERSTUDY - A performer hired to do a role only if the featured player is unable to perform; used primarily in live theatre.

U.B.C.P. - Union of B.C. Performers

U.D.A. - Union des Artistes

UPGRADE - The promotion of an extra performer in a scene to the category of principal performer.

UPM - Unit Production Manager - Oversees the crews and is handles the scheduling and all the technical responsibilities of the production.

UP STAGE - (a) The area located at the back of the stage. Down Stage is the area in front of the performer. (b) To draw attention to oneself at the expense of a fellow performer.

V.O. - Voice over. An off-camera voice coming either from an actor not in the frame, or from a secondary source such as a speakerphone or answering machine.

VOUCHER - Time slip with all pertinent information needed for getting paid properly.

W - A notation on the call sheet indicating that an actor is working that day.

WAIVERS - Union-approved permission for deviation from the terms of a contract.

WALKAWAY - A meal break in which all cast and crew are on their own to get lunch.

WARDROBE - The clothing a performer wears on camera.

WARDROBE ALLOWANCE - A maintenance fee paid to on-camera talent for the use (and dry cleaning) of talent's own clothing.

WARDROBE FITTING - A session held prior to production to prepare a performer's costumes.

WEATHER PERMIT CALL - Due to weather conditions, the production company has the option to release an actor four hours after the call time (if the camera has not started to roll) with a reduced rate of pay for the day.

WEB BROWSER - A browser is an application program that provides a way to look at and interact with all the information on the World Wide Web. Commonly: Internet Explorer (IE), Netscape, Opera, Lynx, Mozilla

WGA - Writers Guild of America

WGC - Writers Guild of Canada

W/N - Will Notify. A notation on a call sheet that tells the actor that he/she will probably work that day but the specific time has not yet been decided.

WRAP - The completion of a day's filming or of the entire production.

WRAP PARTY - The end of the production party.

ZOOM - A camera technique with a special lens to adjust the depth of a shot, accomplished without moving the camera.

WHAT DOES AN AGENT DO?

Your agent who has the most clearly defined role. Your personal manager is responsible for advising you on all facets of your career, from your image to your choice of collaborators. Your attorney is responsible for a wide range of legal issues ranging from setting up your business structure to reviewing your contracts to protecting your intellectual property. The agent has only one function, getting you employment.

Talent agencies range from the giants such as CAA, ICM and William Morris to the smaller one-person operations. While the large agencies cover the whole spectrum, the smaller agencies often specialize in different areas of the entertainment industry such as nightclubs, dance companies, film, corporate parties, etc. Talent agents (also known as booking agents) not surprisingly tend to sign performers who have a demonstrated ability to fill a room with warm bodies. This means that many artists end up doing their own booking initially.

An Agent is a state licensed individual who sends models, actors, singers, etc., to auditions and castings. They are only supposed to make their money off of a percentage of bookings generated by the models and talent they handle. For example, in Florida it is illegal for an agent to make money in any other way except for bookings income. In that state and a few others, they are not allowed to hold classes or photo sessions. In most states there is no such restriction; therefore, many agents make their money off of packaging (photos, printing, etc.) and do very little to book talent. Legitimate agents tend not to take a major interest in directly guiding the careers of models or talent (actors, singers, dancers, and comedians). Their role is to send a number of qualified, packaged prospects to auditions, throwing them up against the proverbial wall and seeing who sticks. Those who stick make money for the agents who get a percentage of the income generated from each job won. If a model or talent comes to the agent unpackaged, or with bad packaging (portfolios, composites, head shots, video reels, audio reels), then the agent cannot do much with them except to send actors out to be extras and models out to do store promotions. If the model comes with a top notch, truly competitive professional portfolio, a good well-printed composite and possibly an industry standard web page, the agent can definitely send her/him to auditions for highly paid print work or (if he or she has the height) for runway opportunities: The same with actors who come with solid training, professional headshots, web pages and video reels.

WHAT DOES A MANAGER DO?

A Manager is the personal mentor and guide for an individual model or talent. Managers, or talent management companies, are among the sources that are intended to be packaging and training resources by such regulated states as Florida. This keeps the services competitive. There is no "if you do these things, I will promise you a job" with a legitimate manager or management company. Managers are more or less the quarterback of the team (model/talent, manager, and agents), setting a direction, telling the model/talent what they need to do to compete -- and giving them the bad news in terms of what they cannot do. Managers are also allowed to make a percentage of what the model or talent earns, not for specific bookings but for their overall earnings generated in the entertainment arena. However, many management companies simply generate their income through their packaging, training and counseling services, and allow the model or actor to keep the income that they earn without sharing it.

WHAT DOES A CASTING DIRECTOR DO?

Casting Directors are the counterparts of the agent. Whereas the agent represents the model/talent at an audition, the casting director represents the client; the film Production Company, commercial production company, ad agency, and catalog house in finding talent. They also do nothing to guide the talent. They just look for the best possible candidates to present to their client. Many casting directors generate side income by teaching occasional acting classes focusing upon the art of the audition.

WHAT DOES A TALENT SCOUT DO?

A Talent Scout works usually for a manager or management group. They can work for an agent but are often more of a voluntary service to the agent. This is due to the fact that, in certain states, scouts are not even allowed to make a percentage of what the agent earns because by state law (as in Florida or Texas), anyone who makes income from a model or talent for a specific job is by definition an agent and must be licensed as an agent. Normally, talent scouts work for management groups, finding new talent in a variety of ways and convincing them to work with the specific management group. Since the management team can make money legally by offering packaging (portfolios, head shots, promo shots, printing, etc.) and training (film technique classes, modeling classes, stunt schools, etc.), the talent scout, in finding a new unpackaged, untrained model or talent is able to make a percentage of the income that the management group makes. This is why ethics is important.

WHAT DOES A TALENT COORDINATOR DO?

A Talent Coordinator also works for a manager or management group in a manner similar to the talent scout. A coordinator is a superset of the scout with additional duties and opportunities. A talent coordinator not only scouts model and talent, but is also will responsible for setting up workshops and photo shoots. The talent coordinator will sponsor and supervise events specifically for talent or models he/she discovers. An event coordinator will take this a step further and will sponsor an event (workshop, conference, model search, talent search or photo shoot) that will include both talent and models he/she has discovered as well as talent or models whom other scouts or coordinators may find - making an override off all of the attendees. The event coordinator, therefore, has the ability to generate earnings off of scouts and coordinators that he/she may not even know. Further, both forms of coordinators are allowed to build their own talent scout teams and generate overrides off of the production of their teams. Talent scouts do not build scout teams. So it could be said that coordinators bear their title because they coordinate both events and talent scouts.

Becoming A Talent Agent

www.StyleCareer.com

Breaking Into and Succeeding as a Talent Agent
Starting a Talent Agency
Representing Local, National and International Talents
Representing Models, Artists, Musicians, and others

Pursue your dream working with celebrities now. If you love the entertainment industry and people say you're a people person and trustworthy, this prestigious career is yours.

Contrary to popular belief, talent agencies are not only located in Los Angeles and New York. They don't only represent A-list celebrities either. Most likely, there is at least one agency in your hometown that represents local talents. Therefore, there is still a place for another one (you).

Many larger-than-life talent agents started small at home, equipped with nothing more than a computer and a phone line, which is the beauty of this business. You can start as small as you want with only hundreds of dollars in starting capital. Or, if you have secured a business loan, you can also rent an office space with prestigious downtown mailing address.

Above all, **a talent agent must be ethical and professional**. Without these traits, it is almost impossible to stay afloat in this field. Word-of-mouth can be good or bad. Most likely, you wouldn't want to experience the latter.

Now, where are the jobs? Depending on where you offer your services, the sky can be the limit. Naturally, if your agency is located in a big city, for instance, it's more likely that the talents get better fees and more jobs. But anywhere you're locate, there must be at least a few job sources:

Modeling industry

Magazines, Catalogs, Runways, Advertisements, Fit modeling, TV shows, Showrooms Exhibitions and Business launches

Acting industry

TV shows, Video clips, Corporate training videos, Broadway shows, Voiceovers and Stand-up comics

Music industry

Bars and clubs, Jazz clubs, Recording studios, Radios, Background music sessions and Concerts

Today, starting a talent agency is pretty simple. With sufficient preparation on what you're going to encounter, you should be able to break into and succeed in this growing field with confidence. **StyleCareer.com Breaking Into and Succeeding as a Talent Agent** contains all the information needed to secure your first talent and job.

If you're still unsure on the kinds of talent to represent in your hometown, consider the following:

Master of Ceremony (MC), Singer, Stand-up Comedian, Illusionist, Clown, Comedian and Mime

Auxiliary services can also be considered to "piggyback" your services. These are basically non-competing items that will complement your existing service, which is securing jobs for your talents.

- Photographers
- Videographers
- Stage Managers
- On-site Producers
- Sound Systems
- Lighting Systems
- Caterers
- Florists
- Security Guards
- Tent Supplies
- Silverware Supplies
- Stage Rentals
- Pony Rides
- Carnival Rides

This **StyleCareer.com eGuide Breaking Into and Succeeding as a Talent Agent** e provides insightful information, advices and tips for anyone who is contemplating to become a talk show host. Numerous hard-to-find resources are included to help you locate pertinent information.

Do you know that there are 2.5 million wedding ceremonies conducted each year in the United States?

This alone translates into US$40 billion spent each year for wedding-related products and services, including catering and entertainment, such as music and festive celebrations.

I'm sure now you'd be able to imagine how lucrative it is being a talent agent in your hometown. Earning anywhere from 10% to 25% of the talent's earning as your fee, your income potential is unlimited.

Editor-in-Chief **Jennie S. Bev** and co-author **Valeria W. Barber** have talked to several talent agents, to provide you with **the best and the most comprehensive self-study guidebook filled with insider information, tips and advice** for breaking into and succeeding as a talent agent.
Among them are:

- **A former actor**
- **A former child actress**
- **Founders of local and national agencies**

Editor-in-Chief **Jennie S. Bev** was named **2003 EPPIE Award** finalist in Non-Fiction How To category for excellence in electronic publishing.

Read about her in the April 2004 issue of ***Entrepreneur***, an international business magazine with 550,000 copies in circulation worldwide. Interviewed by Staff Writer Nichole Torres, she talked about breaking into and succeeding as a fashion entrepreneur.

Jennie's feature as a **success story profile** ran in ***Home Business*** September/October 2004 issue. Interviewed by Staff Writer Sandy Larson, she talked about what motivated her to start **StyleCareer.com** and help people breaking into fashion and image careers.

Co-author **Valeria W. Barber** is an award-winning author who has been published in over 150 magazines, newspapers, and anthologies including ***The Noe Valley Voice*, *Regeneration***, and ***SF Bay Guardian***. She provides web content and informative articles for a variety of businesses and web sites on every topic, as well as transcription and editing services through her business, The T.E.A. Service. She was awarded Notable Story of 2004 by the **Million Writers Award** and holds an MFA in Writing

This **61-page** instantly downloadable **StyleCareer.com eGuide "Breaking Into and Succeeding as a Talent Agent"** brings you valuable insider tips, advice and suggestions **not available elsewhere** to help you embark on this highly lucrative career path.

TABLE OF CONTENTS

This eGuide **literally saves** you **hundreds or even thousands** of dollars spent on **professional workshops and training** and hundreds of hours of research. It will help you break into the career of your dream faster without having to repeat the same mistakes that most newcomers do.

They have also included online resources on the following:

- Sample contracts
- Unions
- Trade associations
- Trends in the industry

CONTRACTS

There exist a variety of degrees, or levels, of contract used in the professional industry to represent and/or handle models, actors, singers and other talent. There are many variations, but primarily three basic levels that include non-exclusive agency representation, exclusive agency representation, and management / personal manager agreements. This article is set up to define these three levels in very simple terms. Each is much more complex than discussed here, so consider this simply an overview that defines and briefly outlines the differences in each, as well as some of the pros and cons of each.

Keep in mind that in many states in the U.S., it is required for each of these situations that the agent and/or manager be licensed through the state in which they operate, and that they hold a bond payable to that state to protect the talent from default in payment.

Non-Exclusive Agency Agreement

Probably the least binding of agency agreements, the non-exclusive simply allows an agency to submit you for projects. It does not lock you to that agency, it does not restrict you from finding your own work, it does not prevent you from working with other agencies, and it does not bind you into terms that keep you with the agency for any specified amount of time.

Is the non-exclusive contract good or bad? It can be either, or both. It all depends upon the agency and you and your needs. It depends upon the agency size and contacts, and the market you are in as well. In solid talent markets, non-exclusive agreements are unheard of. In smaller markets, they are often used more frequently.

Pro's

- You're not locked into an agency for an amount of time;
- You may cancel any time;
- You may find work on your own and not involve the agent;
- You may have multiple agents in the same region.

Con's

- You will receive less attention that someone who's exclusive;
- If you sign with another agent and they dual submit you for projects, chances are you will confuse the casting people and be discarded;
- Your future with the agency is uncertain;
- You are on the books, but not submitted as heavily as someone exclusive.

Dual submission is a major issue with talent signed with multiple agents. If the agents are in different markets, or handle different types of work, this is generally not an issue. If, however, a talent signs with two similar agencies in the same town, the chances are both agencies will receive the same calls from casting people and ad agencies - since most contact from a list of agents to see a variety of people. When both agents submit you for the same project, it hurts the agent and it hurts the talent. A casting person doesn't want to see the same person from two agents - **WHAT'S THE POINT OF USING MORE THAN ONE AGENCY IF THE PEOPLE SUBMITTED ARE THE SAME?** For this reason, agencies handle both exclusive and non-exclusive talent submit differently based upon these agreements and the implications that such submissions may make on potential clients. For instance, if the agent has a non-exclusive with an actor and that actor is also with a neighboring agency... and if a client exclusively is working with the agent then that agent will provide that talent as an option to the client. If, however, there is an open submission call that all the local agencies are submitting for... the agent may be shy about submitting someone that they know their neighbor agent (effectively, their competition) may be submitting also. For this reason, the non-exclusive agreement can be the least desirable for a talent who really wants to optimize their potential through agency representation (because imagine that if both agents follow this same thinking, then that person doesn't get submitted at all!).

Exclusive Agency Agreement

Now enter the next level up in professionalism. This is the binding one, though it may be written in various ways depending upon the needs of the agent. Some exclusive agreements may lock a person in with that agency for one, two, or even three years. Other agencies leave their exclusive open and allow them to cancel at any time, providing proper notice. Exclusively signed talents are worked hard for. The agent has locked them into exclusive because there is more value in them. As such, the agent invests more in time to these people in order to promote them. In medium to large markets, this is the only type of contract offered. Agents do not want to work hard for someone in submitting only to find out they were submitted also by someone else and either discarded, or awarded the role or job through the other agent. Once or twice of this, and the agent will generally forget the person exists, or cancel contract. In the exclusive, the talent is not allowed to sign with other agents. The talent may or may not be due to the agent commission on jobs they find on their own as well. My personal opinion is that if an agent is not involved in finding the work, then the talent should not have to give commission. Many agencies, however, do not believe the same and if the talent lands their own jobs, they still pay their agent a commission. For most aspiring professionals, signing exclusive is a goal above a non-exclusive. To be in an agreement that motivates your agency to work harder for you should be the goal of every talent. If you can find an agency that offers you exclusive representation and does not lock you into a certain amount of time with them, that is the ultimate achievement.

Management (and/or the Personal Manager)

Some agent's act as managers by virtue of the way that they do business. In addition to job finding, they help their talent develop in their goals and their career. These are the type of agents, the ones who take a personal interest in their people that should be most desirable. Strive not to sign and become a number in a file and a headshot on a wall. Know your agent, become friends, work with them and develop a career that they can market. Beyond this, you may find Personal Management the next level of contract. This is separate from an agency agreement in that a Personal Manager does not find you jobs. A manager develops your direction, helps you with achieving your goals, gets you signed with agents who can get you work, and holds your hand through the development and management of your career. Generally, the Personal Manager is someone that the talent already knows and trusts. It is necessary in many states that the manager be licensed the same as an agent, so use caution and make sure if you have management, that they hold the same level of bond and licensing required by your state as for an agency.

Personal Manager's will lock you in for several years, minimum. This is necessary since what they will do for you does not see results immediately. It takes time to develop a person and to market them, and the manager must be allowed that time and not discarded a few months down the road before any opportunity to see results from their efforts. A common term is 3 years, though 5 years is also acceptable practice in the industry. The difference with a manager and an agent, besides the fact that agents find jobs and managers develop you so you land the jobs the agent finds, is that you must have 100% dedication and trust in the person who is managing you. Unlike with an agent where if you disagree with something you can go off and do what you prefer, if you Personal Manager instructs you to do something specific, change your style, alter your direction, etc., then you need to do it. If you do not, then you are not being "managed."

MODEL & TALENT AGREEMENT /Non Exclusive Sample

NAME____(Model)

DATE______________________ S.S.#___________________________

ADDRESS____

CITY _________________ STATE__________ ZIP CODE_______________

In consideration of _______________ (Modeling Agency), placing my name on its Talent List and using reasonable efforts to procure assignments for me as a model or in other phases of modeling and acting as provided herein, and, at my request, advising me with respect to my career:

I hereby appoint AGENCY as my non-exclusive agent to procure Modeling Assignments for me. As used in this Agreement, the term "Modeling Assignments" shall include the performance of services as a model, actor or actress.

I acknowledge and agree that AGENCY currently has in effect Agreements with other persons, that are the same, or similar, to this Agreement, and that AGENCY will, during the term of this Agreement, enter into additional such Agreements, and that AGENCY will provide and furnish to other persons, the same or similar services as AGENCY has undertaken to provide to me under this Agreement.

By its acceptance hereof, AGENCY acknowledges its understanding that I may procure Modeling Assignments for myself or through sources other than AGENCY, and that AGENCY shall not be entitled to any fee, except on Modeling Assignments which AGENCY procures for me.

Unless otherwise mutually agreed to in writing and signed by AGENCY and me, all compensation for my services in the performance of Modeling Assignments procured for me by AGENCY shall be paid directly to AGENCY, and AGENCY shall deduct, as its fee, a sum equal to 20% of the sum paid by the client for whom such services were rendered, before AGENCY forwards payment to me for my services. I further agree that AGENCY shall be entitled to receive such percentage of all compensation payable for my services throughout the term of this Agreement, and any compensation which may become payable in the future for contracts, engagements and/or Agreements procured by AGENCY for me and entered into or negotiated for during the term of this Agreement and on all extensions or renewals of such contracts, engagements and/or Agreements. In the event that any compensation shall be paid by any client directly to me, I agree to immediately deliver 20% of such payment to Agency.

I acknowledge and agree that I am not, nor shall I be, deemed for any purpose to be, an employee of AGENCY. I further acknowledge and agree that, at all times, I shall be acting under this Agreement solely and exclusively as an independent contractor. AGENCY shall not be liable or responsible for any injury or damage suffered by me while I am performing Modeling Assignments, whether obtained directly or indirectly by or through AGENCY, nor will AGENCY maintain any liability, Worker's Compensation, unemployment or other form or type of insurance coverage for or upon me. I hereby accept and agree to be solely responsible for, all risks of performing Modeling Assignments under this Agreement, and I hereby agree to indemnify and hold AGENCY harmless of and from any and all claims, demands, damages, costs and expenses, of any and every kind, which may arise or which may be claimed against AGENCY, arising from my performance of services under this Agreement.

My (our) signature(s) below affirm that both sides of this agreement contain Disclosures. I (we) have read and agree to all terms and conditions as disclosed on both sides of this Agreement.

Signature of Model______________________________

MY SIGNATURE ABOVE CERTIFIES THAT I HAVE FURNISHED THE CORRECT SSN ABOVE, AND THAT I AM NOT SUBJECT TO BACKUP WITHHOLDING.

Accepted: AGENCY

By: ____

Signature of Parent or Legal Guardian (if Model under 18 years)

AGENCY CONTRACT/Samples

Management and agency contracts are substantially similar with only a few distinctions. Most agents will represent artists on an exclusive basis. This means that anyone who wants to engage the artist must negotiate with the agent. This also means that the agency may be entitled to a commission derived from virtually all of artist's employment in the entertainment industry.

Generally, talent agency agreements have an initial term of one to three years. Additional one-year options are possible, sometimes based on certain specified levels of income having been achieved during the term. Agents generally receive a commission which ranges between ten and twenty percent. However, most performers' union agreements impose a 10% cap on agent's commissions. The artist should always limit the commissionable income to live performance monies and exclude any recording and publishing income. Furthermore, unless the artist is signed to one of the big agencies, which put together package deals in a variety of media, the artist will not want the music agent to be commissioning income derived from acting gigs.

Except for the provisions relating to commissions, agency agreements, unlike management agreements, tend to be non-negotiable. One reason for this is that talent agencies that work with union-affiliated artists are obligated to use artist-friendly forms of agreement drafted by the union lawyers. Furthermore, it is important to note that some states, such as California, require that talent agencies be licensed by the State Labor Commission. As a result, in the State of California, a non-licensed individual, including a manager, may not secure employment for an artist. An exception to this law does permit a non-licensed individual from procuring a recording contract for an artist.

The rationale for regulating talent agencies is to protect artists from entering into unfair contracts. California's licensing procedure consists of requiring prospective agents to be bonded and fingerprinted, and for the approval of the talent agencies contracts by the state labor commission. Furthermore, each agent must post in their office a schedule of fees and a description of dispute resolution procedures. In California, the Division of Labor Standards Enforcement handles the approval of contracts. Their mandate is set forth in Labor Code Section 1700.2:

> Every talent agency shall submit to the Labor Commissioner a form or forms of contract to be utilized by such talent agency in entering into written contracts with artists for the employment of the services of such talent agency by such artists, and secure the approval of the Labor Commissioner thereof. Such approval shall not be withheld as to any proposed form of contract unless such proposed form of contract is unfair, unjust and oppressive to the artist.

Significantly, the approval process results in several of the major distinctions between management contracts and talent agency contracts. Consider the following provision, which would be standard in talent agency agreements in every state, including, until recently, California:

> Agent is not required to make any loans or advances hereunder to Artist or for Artist account nor incur any expenses on Artist behalf, but, in the event Agent does so, Artist shall within repay to, or reimburse, Agent no less frequently than monthly for such loans, advances and or expenses incurred (including, without limitation, photocopying and postage expenses, air and ground transportation and lodging and living expenses while traveling, promotion and publicity expenses, conference fees, registration fees, exhibition booth fees, equipment rental fees, showcase expenses, and all other expenses, whether incurred prior to or during the Term, relating to Agent's services in accordance with this Agreement).

The DLSE's current policy is that a talent agency contract may not provide for reimbursement of expenses by the agency. The DLSE view is that a talent agency's sole source of compensation is commissions and that such reimbursement for expenses would be analogous to a prohibited registration fee. Firstly, there is an obvious distinction between a registration fee and the reimbursement of expenses for specific items. For example, if a talent agency should incur expenses in connection with attending an industry conference on behalf of a specific artist, the expenses of travel and registration fees are commonly subject to reimbursement by the agency. Secondly, that argument is contrary to the provisions of Labor Code Section 1700.40 (a), which reads as follows:

> No talent agency shall collect a registration fee. In the event that a talent agency shall collect from an artist a fee or expenses for obtaining employment for the artist, and the artist shall fail to procure the employment, or the artist shall fail to be paid for the employment, the talent agency shall, upon demand therefore, repay to the artist the fee and expenses so collected. Unless repayment thereof is made within 48 hours after demand therefore, the talent agency shall pay to the artist an additional sum equal to the amount of the fee.

Clearly, the express reference to "a fee or expenses" indicates that a contractual provision dealing with "expenses" is customary and legitimate. And there is no section of the Labor Code or the California Code of Regulations, which would prohibit such a provision. As stated above, Section 1700.23 of the Code provides that approval "shall not be withheld as to any proposed form of contract unless such proposed form of contract is unfair, unjust and oppressive to the artist." Accordingly, the DLSE's disapproval of this provision without the support of any express statutory or regulatory authority, or without any legitimate public purpose, is not only unfair to talent agencies, but is also of dubious legal authority. In addition, many talent agency contracts have a provision, which allows the agent to collect a commission for a period of time after the term of the agreement derived from rebooking which originated from the agent's original efforts. Such a provision might state that:

> If, within one (1) year after the end of the Term hereof, Artist accepts any offer on Terms similar or reasonably comparable to any offer made to Artist during the Term hereof, from or through the same offer or any person, firm or corporation directly or indirectly connected with such offer or, the contract resulting there from (oral or written), shall be subject to all the Terms hereof.

Again, to the best of my knowledge, there is no section of the Labor Code or the California Code of Regulations, which would prohibit such a provision. Nevertheless, the DLSE has stated that such a provision will not be approved, unless the period is limited to four months, because it would give an unfair advantage to any one agency. However, this would appear not to be an appropriate basis for disapproval. Section 1700.23 of the Code provides that approval "shall not be withheld as to any proposed form of contract unless such proposed form of contract is unfair, unjust and oppressive to the artist". Competitiveness with other talent agencies is not a legitimate purpose of state's review process.

Finally, unlike most management agreements, indemnification of the agency by the Artist is not permitted. Nor is a provision for injunctive relief in the event of any dispute between artist and agency.

Contract for Managing an Artist Sample

This agreement is made this 1st day of January, 2004 by and between Child Star (hereinafter referred to as the "Manager") and the performer, if more than one, listed on Addendum A attached hereto and included herein (hereinafter referred to as the "Artist"), by and through their designated agent or representative ("Manager") identified below.
WHEREAS, Presenter conducts the event known as: Superstar Kid (hereinafter referred to as the "Artist "),
WHEREAS, Manager desires to Administrative, Agent, and Management duties for Artist, as independent contractor(s), to provide the duties generally described below;
WHEREAS, Manager(s) desire to provide such duties;
The parties agree as follows:
1. Agent/Manager: The name and mailing address of the Representative, who is executing this Agreement on behalf of Artist(s), is: ____
3. Place of Performance: The place of performance is at
____
4. Date(s) and Time(s) of Performance:
The date(s) of the Performance shall be ____________________, 2002 and the time(s) of the Performance shall be ____________________. This Performance shall be _______ hours with a _________ intermission.
5. Performance: The Performance is generally described as:
____
6. Agreement to Perform: Artist(s) agree to provide the Performance in accordance with the terms of this Agreement and any addendums or riders hereto.
7. Price of Performance: Presenter agrees to pay Artist or his agent an aggregate of ____________________ DOLLARS ($0000) for the Performance by cheque immediately following the Performance. The cheque shall be made payable to: ____.
8. Recording, Reproduction or Transmission of Performance: Presenter will use its best efforts to prevent the recording, reproduction or transmission of the Performance without the written permission of Artist(s) or Artist's representative.
9. Excuse of Obligations: Presenter and Artist shall be excused from their obligations hereunder in the event of proven sickness, accident, riot, strike, epidemic, act of God or any other legitimate condition or occurrence beyond their respective control.
10. Taxes: Presenter agrees to prepare and file all tax information required of a person who hires an independent contractor and Artist(s) agree that they have sole responsibility for the payment of any federal or provincial taxes arising from the monies paid by Presenter to Artist(s) for the Performance.
11. Indemnify for Copyright Infringement: Artist(s) represent and warrant that they are knowledgeable about the copyright laws of Canada as applicable to the Performance, and that Artist(s) shall not perform any copyrighted materials of others during Performance without full compliance with such applicable copyright laws. In the event that Artist(s) breach this representation, warranty and covenant, Artist(s) hereby agree to INDEMNIFY AND HOLD HARMLESS Presenter and its employees, guests and agents from and against all liability, loss, damages, claims, and expenses (including attorney's fees) arising out of such breach.
12. Independent Contractor: Artist(s) acknowledge that they shall perform their obligations hereunder as an independent contractor and not as an employee of Presenter. Artist(s) further acknowledge that they are not on Presenter's payroll and/or tax withholding rolls. Artist(s) shall have sole control and direction in the conduct of the Performance.
13. Merchandising: Artist(s) shall not sell any goods, products, merchandise or services (other than the services provided herein) at the performance except by express written permission of Presenter.
14. Promotion: Presenter shall be entitled to advertise and promote the appearance of Artist(s) and the Performance. Artist(s) acknowledge that Presenter will rely on the terms hereof in all such promotions and advertising and in the brochures to be printed setting forth the names, dates and times of all performances to be held. Artist(s) hereby acknowledge and agree that Presenter may use their names, photographs, likeness, facsimile signature and any other promotional materials in all of such promotions, advertising or other activities used to increase attendance at performance.

15. Parking: Presenter shall provide parking space for vehicles in a location of close proximity to and with direct access to the backstage area where Performance will take place on the date(s) of Performance. This parking space will be reserved for Artist(s) for a period of four (4) hours prior to the Performance and ending three (3) hours following the Performance.
16. Security: Presenter shall provide security for the backstage and stage areas before, during and after the Performance. Presenter shall provide security personnel to protect Artist(s) and their property as deemed appropriate by Presenter in its discretion.
17. Passes: Presenter shall provide identification passes to Artist(s) for the backstage and stage where Performance is to be held.
18. Stage: At its sole expense, Presenter shall furnish the stage, and stage lighting, sound and power for the Performance, and Presenter shall provide all stagehands required to assist the set-up for and conduct of the Performance and takedown after the Performance.
19. Dressing Rooms: Presenter shall provide Artist(s) with one private dressing room, which will be clean, dry, well-lit and air-conditioned.
20. Authority to Execute: The representative who is executing this Agreement on behalf of Artist(s) hereby warrants and represents that he has the full power and authority to bind Artist(s) on whose behalf he is executing this Agreement and acknowledges that he is making this representation and warranty with the understanding that Presenter is relying thereon.
In Witness Hereof, this Agreement is executed on the date first above written.
Presenter (Authorized signature)
By: ______________________________
Date: ______________________________
Artist/Agent or Manager Representative
By: ______________________________
Date: ______________________________

POPULAR ENTERTAINMENT LAWYERS

www.barryoliverchase.com

21 SE First Ave. Suite 700 Miami, Florida 33131 Phone: (305) 373-7665 Fax: (305) 373-7668

Welcome to the **Law Office of BARRY OLIVE CHASE**, an **Entertainment, Arts, Media and Sports** - related legal practice. From its office in **Miami, Florida**, the firm provides legal advice and representation to national and international entertainment, arts, sports and media industry clients. The firm provides a full gamut of legal services to clients in these industries, including:

- Television, Film, Music, Radio & Live Performances (including adult entertainment, magicians, comedy acts), Modeling and Theater
- Intellectual Property: Copyright, Trademark and Celebrity Endorsements/Licensing/Merchandising
- Literary Property and Publishing
- Sports and Related Merchandising and Licensing
- Internet Law

Barry Oliver Chase has a unique combination of education and experience that sets him apart from most other entertainment attorneys. As a well-trained thinker (honors degrees from both Yale and Harvard) with an insider's perspective on the entertainment industry, Mr. Chase provides his clients with legal representation at the highest level.

Thank you for visiting the website of the **Law Office of Barry Oliver Chase**. Please use this site to learn more about the firm, its practice, and how it can help you with your entertainment, arts, sports and media-related legal needs. You may select an area of interest from the buttons above, or feel free to contact Mr. Chase directly at (305) 373-7665. If you prefer, you may e-mail the firm at LOBOC@aol.com.

Mitchell Graham & Stroud PC

Wachovia Bank Building
Suite 919
315 W. Ponce de Leon Ave.
Decatur, GA 30030-2441

(404) 378-3313/Fax (404) 378-2278

555 Armour Circle
Atlanta, GA 30324-4003

(404) 892-6602/Fax (404) 892-6353

Attorney Sonya B. Stephens, Entertainment

FRASER - ENTERTAINMENT LAW/www.frazer-elaw.com

Canada

110 Spadina Avenue
Suite 222

Toronto, Ontario M5V 2K4
Tel: (416) 967-0080
Fax: (416) 967-0090

United States

9595 Wilshire Boulevard
Suite 900
Beverly Hills, California 90212
Tel: (310) 246-1867
Fax: (310) 246-1866

Email: info@fraser-elaw.com

Fraser – Entertainment Law focuses on serving the needs of clients in the film, television, music and internet industries in Canada and the United States.

Entertainment and Media

The entertainment and media industries enjoy a "fun" and "sexy" image. Popular entertainment is preoccupied with beautiful stars and rich imagery. However, the people who create and develop such content are keenly aware that behind the scenes the work involved in producing compelling material is not always glamorous.

Fraser – Entertainment Law assists its clients in navigating the constantly changing media and entertainment industry environments that they work in.

Entertainment Lawyers at Zuber & Taillieu LLP

Contract negotiations in the film, television, and music industries touch on a large number of complex legal issues that require the advice of skilled, knowledgeable and experienced attorneys. Zuber & Taillieu LLP has cultivated the requisite industry knowledge to effectively represent composers, actors, talk show hosts, production companies, writers, and directors. Our entertainment lawyers review production agreements in all areas of the entertainment industry and protect our clients' interests by carefully evaluating contract terms before any signing. Our law firm represents both prominent and lesser known artists, as well as both plaintiffs and defendants in litigation and arbitration matters. In select circumstances, Zuber & Taillieu LLP also acts as a Producer's Rep for completed feature films of exceptional commercial and artistic merit.

With offices in Beverly Hills and Irvine, California, we represent clients throughout the world. Contact an entertainment lawyer at Zuber & Taillieu LLP today. We offer a free initial consultation, and are available for even weekend and evening appointments.

Entertainment Law

The entertainment attorneys at Zuber & Taillieu LLP provide expert legal counsel to clients regarding:

- Film and TV Production Contracts
- Movie and Television Development and Distribution Contracts
- Music Industry Agreements
- Entertainment Business Entity Formation Filings
- Financing Agreements

- Representation Agreements

Bolster Your Negotiating Position

Don't wait for second chances to achieve your most ambitious goals. Shifts in cultural trends and demographics often determine overnight success or failure. If you aren't prepared to act when an opportunity arises, others in the industry will be happy to step forward. In Hollywood, taking advantage of an opportunity often means effectively negotiating the terms of a contract in a timely manner. We'll negotiate and draft terms that strengthen your position as an artist and secure the rights and protections you need as a business person.

Commitment and Resolve

No one wants to give away what they don't have to. In the world of entertainment, studios, recording companies, as well as the artists themselves prefer to maintain as much control over a product as possible. The entertainment lawyers at Zuber & Taillieu LLP excel at the art of negotiation, and we are fully prepared to aggressively defend your interests.

As your legal counsel, we'll stand by your side, and will not be intimidated. We're familiar with tactics opposing parties will try to use to garner negotiating leverage. In the end, though, it is up to you alone to determine whether or not a deal is acceptable. Zuber & Taillieu LLP will ensure you have the knowledge and advice you need to evaluate your options.

Contact the Entertainment Law Offices of Zuber & Taillieu LLP

If you are considering an entertainment contract, or have been accused of breach of an entertainment contract, contact an entertainment attorney at Zuber & Taillieu LLP today to learn more about our practice. We offer a free initial consultation, and are available for even weekend and evening appointments. With offices in Beverly Hills and Irvine, we represent clients throughout Southern California, and throughout the world.

Los Angeles County Offices:

Zuber & Taillieu LLP
9595 Wilshire Blvd.
9th Floor
Beverly Hills, CA 90212
Phone: (310) 300-8480
Fax: (310) 300-8481

Orange County Offices

Zuber & Taillieu LLP
2030 Main St.
Ste. 1300
Irvine, CA 92614
Phone: (949) 260-4780
Fax: (949) 260-4781

KIA LAW FIRM, LLC

www.kialawfirm.com

IN NEW YORK

432 Park Avenue South
Second Floor
New York, New York 10016
Tel: 212 684 1001 Ext. 619
Fax: 212 684 0920

IN PARIS

89 Avenue Henri Martin
75116 Paris – France
Tel: 011-33 6 62 44 57 96
Fax: 011-33 1 45 04 57 96

E-MAIL
info@kialawfirm.com

KLF Law Practice was launched by Atossa Kia in 2003 in an effort to bring New York and Paris together, to provide the best and most affordable legal services to clients in both America and Europe. KLF offers its legal services to a wide range of clients who are either seeking to create a legal business structure in New York or Paris, prospecting new markets and promotional opportunities, or simply looking to establish an initial contact in both cities in order to develop their business on a local and international level.

As a small, flexible, dynamic, international structure with extensive legal experience and an array of contacts in the worlds of both the music and visual arts industries, KLF will guide and assist you on the path to fulfill all of your creative and business endeavors both in the U.S. and Europe.

Where are we located?

Based in New York, KLF also maintains an office in Paris.

What are our domains of expertise?

Music and visual arts (e.g. photography, graphic design, and etc.).

Who are our clients?

KLF represents a variety of American and European clients including independent music labels and publishing companies, musicians, producers, DJs, as well as photographers and graphic designers.

For more information or to schedule an appointment, do not hesitate to contact us.

Entertainment Law Firm | Schroder Fidlow PLC

www.schroderfidlow.net

Schroder Fidlow, PLC, is an entertainment law firm located in Richmond, Virginia while serving clients in Los Angeles, New York and throughout world. Each of us, Kirk T. Schroder and Bennett J. Fidlow, has more than 16 years of experience providing legal services to a great variety of clients in the entertainment industry, including major studios, producers of films and TV programming, writers, musicians, and many others.

Los Angeles | New York City

Many of our entertainment clients, past and present, have been based in Los Angeles and New York, but we work anywhere in the world, including foreign film locations such as Canada and Mexico. We have worked as entertainment attorneys for clients such as Sony Pictures, Universal Pictures, MGM/UA, Warner Bros. TV, Fox Family, USA Cable, Universal Home Entertainment Productions, and Premiere Radio, as well as individual authors, artists and independent producers.

Entertainment law firms abound in L.A. and NYC, but few exist with the expertise, experience, and connections that we have accumulated over the years, and even fewer provide the level of quality, personal service and cost-effectiveness that we offer. Our experience as long-time entertainment attorneys, combined with our location, give us the freedom to provide big-city, quality legal work for the entertainment industry at more affordable, small-city rates.

Contact us for a free consultation on your entertainment legal issues, and let us put our years of entertainment law experience to work for you.

Schroder Fidlow, PLC
1901 East Franklin Street
Suite 107
Richmond, VA 23223
Tel: 804-225-0505
Fax: 804-225-0507

DAVID P. CUDNOWSKI

www.talentlawfirm.com

330 N. Wabash Avenue
Suite 2604
Chicago, IL 60611

(312) 595-1040/ Fax (312) 595-1041

The Law Office of David P. Cudnowski, Ltd.
Entertainment Law Attorney
Serving the Greater Chicago Area

Do you work as an actor, writer or director in the greater **Chicago** area? If so, consider working with David P. Cudnowski, Ltd. for your legal and tax service needs.

At some point you may need the services of a lawyer familiar with the needs and nuances of people working in the entertainment field. At the law office of David P. Cudnowski, Ltd. we specialize in providing legal and tax advice and representation to people working in the entertainment and performing arts fields throughout the greater **Chicago, IL** area.

Mr. Cudnowski has worked with hundreds of clients involved in the entertainment industry for almost 20 years.

Since 1986, **David P. Cudnowski** has enjoyed a special relationship with actors, entertainers, writers, and producers all over Chicago.

If you work in the entertainment field and need any of the following services:

- Contract review or negotiation,

- To incorporate, whether a C, S or LLC is desired,
- Tax advice or representation before the Internal Revenue Service, whether on audit, at the Appeals Division or in the United States District Court or United States Tax Court, or
- Trademark or copyright registration.

ENTERTAINMENT CHILDREN BANK ACCOUNTS

www.firstent.org

Beginning January 1, 2004 any minor working on a set must have a true and accurate Statement of Trustee from the minor child's Coogan Account stapled to their work permit. The work permit will be valid for only 10 business days unless the Statement of Trustee is attached. The new law also specifies the requirement that the Coogan Account be in a California bank.

For those who already have a Coogan account, you will remember that when you opened the account you were given some type of paperwork. This paperwork varies from bank to bank. What you need to have is a Trustee Statement - This is YOUR statement, not the banks. The bank is not the trustee of your child's money, YOU (and/or your spouse) are. It has been helpful that most banks provide you with this form, but it is important that you realize it is your trustee statement and you are ultimately responsible for the information on it. What this means is, this is you, signing under penalty of perjury that this account has been set up for your child, that it is a blocked trust, and that all information is accurate and current.

The law indicates that the following information should be on a Trustee Statement:
Name, Address & Phone # of Financial Institution
Name of Minor (beneficiary)
Account Name
Account Number
Trustee Name
Signed under penalty of perjury by trustee

NOTE: Photocopy is Acceptable

Some points to consider: The term Coogan account is widely used, but many of the financial institutions don't use the word Coogan when describing or 'naming' their accounts. You might want to include that wording somewhere on your Statement of Trustee. The studio teachers may be looking for the work Coogan on your statement of trustee. If your financial institution has an out of state address for processing Coogan deposits, you might wish to add a local (in California) branch address to your Statement of Trustee. Again, that will decrease the chances that any teacher will question the appropriateness of your account and your Statement of Trustee.

Please be aware that there are penalties for perjury and you should not put inaccurate information on your Statement of Trustee.

Again, the law also requires that this Trustee Statement be attached (stapled) to the back of a work permit, for the permit to remain valid for 6 months. Without a Trustee Statement attached, it will be void in 10 days. The Labor Commissioner will only issue a work permit to a child once every six months. So if your child's work permit is deemed invalid after 10 days it was issued because it does not have a Statement of Trustee stapled to it, you will not be able to get a new one for 6 months after the work permit was initially issued.

As this new law is put into effect there will be, most likely, some confusion with all parties involved. Some studio teachers may not be aware of the new law in the beginning. There is no need to make an issue of this. Just always be sure that YOU are aware and have the appropriate Statement of Trustee stapled to your Childs work permit.

Back to the Statement of Trustee... does it look like an official document from your bank? You might ask to have it printed on their letterhead. Or, you could cut and paste their letterhead onto your document and make copies. Just be sure the information is correct.

The second form you need is the Deposit Instructions. Some financial institutions may combine the two documents, which are fine. You will need to provide the Deposit Instructions to the 2nd AD or whoever is handling the payroll paperwork on the set.

Deposit instructions should include the following information:

Financial Institution Logo
Name of Account
Name of Minor
Name of Trustee
Routing Number
Social Security Number
Financial Institution contact information

Some points to consider: This is the document that will allow the employer to make a deposit into you child's Coogan account. This information is CRITICAL to the process and must be accurate and complete. If your bank changes account numbers (via merging with another bank) you must remember to be certain you get updated documents from them that accurately indicate the routing number and account number of your child's account. Often an employer will have their own form requesting much of the information already on your Deposit Instructions. They may have a standardized payroll document and will have you enter the information onto their form. If you are presented with that request, certainly fill it out for them, and be accurate with the numbers. Encourage them to also keep a copy of your Deposit instructions.

It is important to keep in mind that there are now two separate Coogan issues at the workplace. 1 - attached to work permit as proof of an existing account 2 - to the payroll entity so that the deposits can be made into the account.

Again, please note: The DLSE will only issue a work permit once every 6 months, per this law. This is their response to someone just continually getting a new permit reissued after the 10 day time period is up. After talking to that office, it appears they are trying to tighten their system.

Please check your child's Coogan paperwork for the specific information listed above. Be sure you can comply with the law and keep their work permit valid for 6 months.

SAMPLE COOGAN CHILDREN ACCOUNT APPLICATION

COVER SHEET

Thank you for contacting First Entertainment Credit Union to set up a Coogan Account.

Below you will find A Coogan Account Signature card (Form 965a), which must be completed, signed (wet signature), and returned.

For our files, we ask that you send us the following:

1. A copy of your child's birth certificate.

2. A copy of your State ID or Driver's License.

3. Original Signature card (Form 965a) signed and dated.

Please send to:

First Entertainment Credit Union
Attn: Gale Miller
6735 Forest Lawn Drive
Hollywood, Ca 90068

- **Please note, if a deposit is not made within 90 days, the account will be closed.**

Sincerely yours,

Gale Miller **Coogan Account Specialist gmill@firstent.org (888) 800-3328 x4369**

NEW ACCOUNT INFORMATION
FOR A COOGAN-MINOR BLOCKED ACCOUNT
MINOR'S INFORMATION

Minor's Name: Social Security #
Address:
Telephone: Mother's Maiden Name:
Date of Birth: Place of Birth:
Referred by/Eligibility:

TRUSTEE INFORMATION

Trustee's Name:
Address:
Home Phone: Business Phone:
Employer's Name:
Email:
Date of Birth: Place of Birth:
Social Security # Mother's Maiden Name:
Driver's License # Drivers License State:

OFFICE USE ONLY

Account # Account subject to Court Order: Yes No

A certified copy of the minor's birth certificate is required before funds can be withdrawn when the beneficiary turns eighteen (18) unless otherwise provided by Court Order. Funds may be transferred to another qualifying federally insured trust account. Written documentation of said account must be provided to the credit union and payment must be made directly to said account.

We reserve the right to change the fees and terms of our accounts at any time without notice. Accounts are subject to current terms and conditions and fees. To that end the following documents have been provided to the trustee:

Receipt of which is hereby acknowledged:

____
Trustee's Signature

____
Date

Form 965a

Contact Information

Main Office
6735 Forest Lawn Drive
Hollywood, CA 90068

Mailing Address
P.O. Box 100
Hollywood, CA 90078-0100

Phone:
(323) 851-3673
(888) 800-3328

Fax:
(323) 851-0383

Warner Bros. Studios*
Building 19

Fax: (818) 954-6679

Branch Manager:
Athena Nowlin

Las Vegas
3100 W. Sahara Ave. #206
Las Vegas, NV 89102

Fax: (702) 251-0432

Branch Manager:
John Drake

New York City
555 Madison Ave., 9th Fl.
New York, NY 10022
Ph: (212) 223-9792
Fax: (212) 223-9793

Branch Manager:
Philip Angelastro

Coogan Minor-Blocked Accounts

Thanks to California's Coogan Act (SB 1162), every time young performers work under an entertainment contract, **15% of the gross earnings are set aside** for them until they reach legal majority. These earnings are the legal property of the minor.

Coogan Minor-Blocked Accounts insure that child actor and athletes receive this portion of their earnings when they reach majority age, now 18. Safely insured, **First Entertainment Coogan Accounts** are accounts that are easily set up to directly accept these funds. Here are some basic facts about this account ...

- The minor is the Beneficiary of the Coogan account. The parent is the Trustee.
- No withdrawals are allowed until the minor turns 18. However, the Trustee may transfer funds to an equivalent account before that time.
- A court order is no longer required to open a Coogan Account.

ASSOCIATIONS

www.agentassociation.com

BENEFITS OF MEMBERSHIP IN ATA:

Named Bargaining Partner for Guild Agency Regulations (WGA, DGA, AFTRA, & expired SAG)

Interpret Agency Regulations

Actor's Agent Search website – averaging 35,000 hits a week

Settlement of Agency Commission Disputes

Testimony and/or Present At Agency Arbitrations

ATA forged alliances – this year's passage of the American Jobs Creation Tax Bill

Informative ATA Newsletters on Website

Round table discussions with your peers

Monthly Board Meetings

Educational Seminars (e.g., Understanding Collective Bargaining Agreements, Do's & Don'ts of Employment Contracts)

ATA/SAG Residual Tracking System

Membership discounts

ATA/CSA (Casting Society of America) Cooperative Committee

ATA/CCDA (Commercial Casting Director Association) Cooperative Committee

Legislation

ATA Full Time State Lobbyist

Minor's representation legal authorities

Entertainment Health Care Insurance References

WGA & DGA Outreach Programs

Membership Meetings

ATA Educational Contract Booklets

Meetings with State Assembly Persons

Meetings with Labor Commissioner and Legal Staff

ATA – OVER 65 YEARS STRONG

THE ASSOCIATION OF TALENT AGENTS (ATA) is a Los Angeles based nonprofit trade association comprised of over 100 talent agencies primarily located in Los Angeles and New York. ATA's premiere membership includes agencies of all sizes.

CASTING DIRECTORS/TELEVISION/FILM

www.castingsociety.com

"We are diplomats; we are counselors; we are negotiators. We are teachers; we are artists and visionaries. We are a key element in the creation and success of every theatre, film and television project ever made..." Mary V. Buck, former CSA President

The Casting Society of America is the premier organization of theatrical Casting Directors in film, television, and theatre. Although it is not a union, its members are a united professional society that consistently set the level of professionalism in casting on which the entertainment industry has come to rely. Its more than 350 members are represented not only in the United States, but also in Canada, England, Australia and Italy.

Screen Actors Guild (SAG): **sag.org**

American Federation of Television and Radio Artists (AFTRA): **aftra.org**

Actors Equity Association (AEA): **actorsequity.org**

Academy Players Directory: **acadpd.org**

Association of Talent Agents (ATA):**agentassociation.com**

Talent Managers Association (TMA): **talentmanagers.org**

Internet Movie Data Base (**IMDB.com**)

Internet Broadway Data Base (**IBDB.com**)

Breakdown Services & Actors Access: **breakdownservices.com**

Casting Workbook: **castingworkbook.com**

LA Casting: **lacasting.com**

Now Casting: **nowcasting.com**

CASTING DIRECTORS' GUILD OF GREAT BRITAIN/www.tcdg.co.uk

The Guild is a professional organization of Casting Directors in the film, television, theatre and commercials communities in the UK who have joined together to further their common interests in establishing a recognized standard of professionalism in the industry, enhancing the stature of the profession, providing a free exchange of information and ideas, honoring the achievements of members and standardization of working practices within the industry.

www.productionhub.com

Artist Management   Casting Directors   Dialog / Dialect Coaches   Dramatic Instruction   Extras Agencies
Headshots / Composites   Models Agencies / Agents   Talent Agencies   Casting Assistants   Casting Facilities
Child Agencies   Talent Agencies

CELEBRITY AGENTS

www.agentassociation.com

Following is a list of some of the celebrities you may find at this website.

BLAIR UNDERWOOD
William Morris Agency • Celebrity Commercials (LA/NY), Commercial (LA/NY), TV/Feature Film (LA/NY), Voice-Over Commercial (LA/NY)
ph. (310) 859-4000

HARRISON FORD
United Talent Agency • TV/Feature Film (LA/NY)
ph. (310) 273-6700

MERYL STREEP
CAA • TV/Feature Film (LA/NY)
ph. (310) 288-4545

MATTHEW MCCONAUGHEY
William Morris Agency • Celebrity Commercials (LA)
ph. (310) 859-4000

CAA • Celebrity Commercials (LA/NY), Commercial (LA/NY), Equity (LA/NY), Internet (LA/NY), Print/Model (LA/NY), Sports/Stunts (LA/NY), TV/Feature Film (LA/NY), Voice-Over Commercial (LA/NY), Voice-Over Theatrical (LA/NY)
ph. (310) 288-4545

DENZEL WASHINGTON
ICM • Equity (LA/NY), Internet (LA/NY), TV/Feature Film (LA/NY), Voice-Over Theatrical (LA/NY)
ph. (310) 550-4000

OPRAH WINFREY
CAA • TV/Feature Film (LA)
ph. (310) 288-4545

GEORGE CLOONEY
CAA • TV/Feature Film (LA)
ph. (310) 288-4545

LINDSAY LOHAN
CAA • TV/Feature Film (LA/NY)
ph. (310) 288-4545

TOM CRUISE
CAA • TV/Feature Film (LA)
ph. (310) 288-4545

MARTIN LAWRENCE
United Talent Agency • TV/Feature Film (LA/NY)
ph. (310) 273-6700

WILL SMITH
CAA • TV/Feature Film (LA)
ph. (310) 288-4545

NICK CANNON
Endeavor • TV/Feature Film (LA/NY)
ph. (310) 248-2000

JENNIFER GARNER
William Morris Agency • Commercial (LA/NY), Voice-Over Commercial (LA/NY)
ph. (310) 859-4000

Endeavor • TV/Feature Film (LA/NY)
ph. (310) 248-2000

JENNIFER ANISTON
CAA • TV/Feature Film (LA)
ph. (310) 288-4545

JAMIE FOXX
CAA • TV/Feature Film (LA)
ph. (310) 288-4545

TYLER PERRY
William Morris Agency • Celebrity Commercials (LA/NY), Commercial (LA/NY), TV/Feature Film (LA/NY), Voice-Over Commercial (LA/NY)
ph. (310) 859-4000

MORRIS CHESTNUT
CAA • TV/Feature Film (LA)
ph. (310) 288-4545

ANTONIO BANDERAS
CAA • TV/Feature Film (LA)
ph. (310) 288-4545

GABRIELLE UNION
William Morris Agency • Celebrity Commercials (LA/NY), Commercial (LA/NY), TV/Feature Film (NY), Voice-Over Commercial (LA/NY)
ph. (310) 859-4000

United Talent Agency • TV/Feature Film (LA)
ph. (310) 273-6700

JANET JACKSON
CAA • TV/Feature Film (LA/NY)
ph. (310) 288-4545

EVA LONGORIA
CAA • Celebrity Commercials (LA/NY), Commercial (LA/NY), Equity (LA/NY), Internet (LA/NY), Print/Model (LA/NY), Sports/Stunts (LA/NY), TV/Feature Film (LA/NY), Voice-Over Commercial (LA/NY), Voice-Over Theatrical (LA/NY)
ph. (310) 288-4545

LUCY LIU
William Morris Agency • Celebrity Commercials (LA/NY), Commercial (LA/NY), Voice-Over Commercial (LA/NY)
ph. (310) 859-4000

NIA LONG
Paradigm • TV/Feature Film (LA)
ph. (310) 277-4400

CHRIS ROCK
Endeavor • TV/Feature Film (LA/NY)
ph. (310) 248-2000

TRACIE THOMAS
The Gersh Agency • TV/Feature Film (LA/NY)
ph. (310) 274-6611
The Gersh Agency
232 N. Canon Drive
Beverly Hills, CA 90210
(310) 274-6611

DUANE MARTIN
Paul Kohner Inc. • Celebrity Commercials (LA/NY), Commercial (LA/NY), Equity (LA/NY), Internet (LA/NY), Print/Model (LA/NY), TV/Feature Film (LA/NY), Voice-Over Commercial (LA/NY), Voice-Over Theatrical (LA/NY)
ph. (310) 550-1060

BRAD PITT
CAA • TV/Feature Film (LA)
ph. (310) 288-4545

VIVICA FOX
ICM • Voice-Over Theatrical (LA/NY)
ph. (310) 550-4000

Paradigm • TV/Feature Film (LA/NY)
ph. (310) 277-4400

JENNIFER LOPEZ
Endeavor • TV/Feature Film (LA/NY)
ph. (310) 248-2000

MARIO VAN PEEBLES
Innovative Artists • Commercial (LA/NY), Equity (LA/NY), TV/Feature Film (LA/NY), Voice-Over Commercial (LA/NY), Voice-Over Theatrical (LA/NY)
ph. (310) 656-0400

FOREST WHITAKER
CAA • TV/Feature Film (LA)
ph. (310) 288-4545

Imperium 7 • Voice-Over Commercial (LA) , Voice-Over Theatrical (LA)
ph. (310) 203-9009

SANAA LATHAN
William Morris Agency • Celebrity Commercials (LA/NY), Commercial (LA/NY), TV/Feature Film (LA/NY), Voice-Over Commercial (LA/NY)
ph. (310) 859-4000

WESLEY SNIPES
United Talent Agency • TV/Feature Film (LA/NY)
ph. (310) 273-6700

KENYA MOORE
Cunningham.Escott.Slevin.Doherty • Commercial (LA/NY)
ph. (310) 475-2111

Coast to Coast • TV/Feature Film (NY)
ph. (323) 845-9200

SMS • Equity (LA), TV/Feature Film (LA)
ph. (310) 289-0909

HILARY DUFF
CAA • TV/Feature Film (LA/NY)
ph. (310) 288-4545

William Morris Agency • Celebrity Commercials (LA/NY), Commercial (LA/NY), Voice-Over Commercial (LA/NY)
ph. (310) 859-4000

JOY BRYANT
CAA • TV/Feature Film (LA)
ph. (310) 288-4545

ASHTON KUTCHER
Endeavor • TV/Feature Film (LA/NY)
ph. (310) 248-2000

RAVEN SYMONE
Ford Model, Inc.
www.fordmodels.com
ph. (212)219-6500

JESSICA ALBA
Endeavor • TV/Feature Film (LA/NY)
ph. (310) 248-2000

ASHLEY OLSEN
United Talent Agency • Celebrity Commercials (LA/NY), Commercial (LA/NY), TV/Feature Film (LA/NY), Voice-Over Commercial (LA/NY), Voice-Over Theatrical (LA/NY)
ph. (310) 273-6700

KATIE HOLMES
CAA • TV/Feature Film (LA)
ph. (310) 288-4545

BEYONCE KNOWLES
CAA • TV/Feature Film (LA)
ph. (310) 288-4545

KYLA PRATT
Cunningham.Escott.Slevin.Doherty • Celebrity Commercials (LA/NY), Commercial (LA/NY), Voice-Over Commercial (LA/NY)
ph. (310) 475-2111

CHRISTINA MILIAN
Innovative Artists • Equity (LA/NY), TV/Feature Film (LA/NY)
ph. (310) 656-0400

BRANDY NORWOOD
William Morris Agency • TV/Feature Film (LA/NY)
ph. (310) 859-4000

KELLY ROWLAND
CAA • TV/Feature Film (LA)
ph. (310) 288-4545

BECOME A FILMMAKER

www.nyfa.com

The New York Film Academy is a film school designed for a new generation of filmmakers who want to learn by making their own projects in a hands-on, intensive program. Traditional film schools often delay the moment when students get their hands-on the cameras and actually make movies. At the Academy you will learn by doing from day one. Whether you enroll in a four-week filmmaking workshop or our one-year film-directing program you will write, shoot, direct, and edit your own short films. Your teachers will demystify the craft of film production and challenge you to master the art of visual storytelling.

Learning to be a filmmaker is not done "by the numbers." It is done by immersion in the process and by knowing the "whys" that teaches the "hows." It is done by hands-on experience, trial and error, feedback and correction, and questions and answers. In the creation of our film school and curriculum, we worked with students and faculty from around the world. We adapted the best ideas and elements from the top university film schools and designed an alternative which provides a thorough grounding in filmmaking, and is also flexible in meeting the diverse needs of our students.

We teach the art and craft of filmmaking through a balance of classroom instruction, hands-on film workshops, and immediate directing experience. Students are behind the camera from the first day of class. Students will learn about and explore the numerous visual, dramatic, and technical challenges that directors face. In addition to writing, producing, directing, and editing their own films, students assist other members of their crews in the roles of director of photography, assistant cameraperson, and gaffer/grip, providing everyone with extensive set experience. Ultimately, through intensive hands-on instruction, our goal is to empower students to artfully, cinematically, and originally tell stories.

Another wonderful aspect of our film school is the student body. The Film Academy draws an incredibly diverse, international group of film students who share a passion for telling stories with moving images. One recent film school class included students from the United States, Canada, France, Italy, England, Switzerland, Portugal, Spain, Australia, Singapore, Mexico, Brazil, India, Israel, Japan, and China.

The combination of non-stop collaborative work, diverse cultures and experiences, and a shared passion for film make our workshops and programs intensely enriching experiences. Students forge bonds, both professional and personal, that often extend beyond the film school. Students often say that the New York Film Academy's workshop or program was the most exciting, difficult, and rewarding creative experience they have ever had.

At the end of each of the Film Academy film courses, the final films are celebrated in a school screening open to cast, crew, friends and family. All students who successfully complete the workshop will receive a New York Film Academy Diploma and leave the film school with a digital master tape of all their projects.

Other New York Film Academy programs and workshops:
Acting for Film & Television | Screenwriting | Digital Filmmaking | Animation | Final Cut Pro Editing | Film and Television Producing | Summer Camps for Teens

NEW YORK HEADQUARTERS
100 E. 17th St., New York, NY 10003
p: 212-674-4300
f: 212-477.1414
email: film@nyfa.com

LOS ANGELES UNIVERSAL STUDIOS
100 Universal City Plaza Drive Bldg. 9128, Universal City, CA
p: 818-733-2600
f: 818-733.4074
email: studios@nyfa.com

ENGLAND
St. Catherine's College in the University of Oxford
Manor Road Oxford OX1 3UJ, United Kingdom
p: +44 1865 271 805
f: +44 1865-271-807
email: filmuk@nyfa.com

FILMMAKING RESOURCES FOR INDEPENDENT FILMMAKERS

FILMMAKER NEWS & GOSSIP
Top 10 Filmmaker Sites

Filmmaking news for filmmakers, screenwriters and directors

Independent Gossip - www.Indiewire.com
Filmmaker Gossip - www.AintItCoolNews.com
Moviemaking News - www.MovieMaker.com
Filmmaking News - www.FilmmakerMagazine.com
Low Budget Film News - www.FilmThreat.com
Indie Film News - www.FadeInMag.com
Independent Film News - www.Filmmaker.com
Film & Movie Gossip - www.RumourMachine.com
Canadian Film News - www.PraxisFilm.com
Independent Film News - www.IndieTalk.com

HOLLYWOOD NEWS & GOSSIP, B.O.
Ten Best Hollywood Sites

Movie Studio & Theatrical Distributors News: Who's Producing & Directing Movies? What Movies Make?

Film Studio News - www.Variety.com
Movie Studio News - www.HollywoodReporter.com
Hollywood Gossip - www.Eonline.com/Gossip
Movie Business News - www.CNN.com/ShowBiz
Hollywood Box Office - www.BoxOfficeMojo.com
Movie Studio Gossip - www.Premiere.com
Top 100 American Films - www.Filmsite.org/BoxOffice
Movie Star Gossip - www.RadarOnline.com
Worldwide Box Office - www.BoxOfficeGuru.com
Global Movie Box Office - www.WorlwideBoxoffice.com

SCREENWRITER NEWS & GOSSIP
Best 10 Screenwriter Sites

Screenwriting secrets, screenplay tricks & script selling information

Writers Guild Magazine - www.WGA.org/Written
Script Sales & News - www.ScriptMag.com
Screenwriting Tricks - www.CreativeScreenwriting.com
Global Screenwriting - www.ScreenTalk.org
New York Screenwriter - www.NYScreenwriter.com
Hollywood Screenwriter - www.HollywoodScriptwriter.com
Writing Movies - www.ScreenTalk.Biz
Screenwriter Magazine - www.ScreenwriterMagazine.com
Screenwriting News - www.ScreenwritersMonthly.com
Online Writers Magazine - www.ScreenwritersUtopia.com

ACTOR NEWS, GOSSIP & UNIONS
10 Best Actor Sites

Inside information on Film Actors, Movie Acting and TV Actors: The Craft of Acting and The Acting Business.

Backstage Newspaper - www.Backstage.com
Film Acting Parts - www.BreakdownServices.com
Hollywood Actor News - www.HollywoodActor.com
Chicago Actor News - www.Performink.com
American Movie Actors Guild - www.SAG.org
English Actors Guild - www.ActorsEquity.org
Theater and Stage Actors - www.AFTRA.org
Australian Movie Actors - www.Alliance.org.au
Canadian Film & TV Actors - www.ACTRA.ca
French Film & Movie Actors - www.UnionDesArtistes.com

INDEPENDENT FILMMAKING BASICS

Top 10 Filmmaking Sites

- Independent Filmmaker Resources. Be a Movie Producer. Be a Film Director: Filmmaking Information.

Filmmaking info - Filmmaking.net
Independent Filmmaker - Filmmaker.com
Filmmaking Encyclopedia - CyberFilmSchool.com
Film Production News - FilmandVideoMagazine.com
Film Editing News - PostMagazine.com
Commercial Filmmaking News - Millimeter.com
Video Equipment News - VideoSystems.com
Film & Video News - IndieSlate.com
Filmmaking Magazine - FilmmakerMagazine.com
Moviemaking Magazine - Moviemaker.com

FILMMAKING & MOVIEMAKING AWARDS

Ten Best Movie Making Sites

Win an Emmy, an Oscar or an Independent Film Award: Filmmaking Awards. Be an Award-winning Filmmaker!

Win an Oscar - Oscar.org
Win an Emmy - Emmys.org
Win a Documentary Award - Documentary.org
Independent Film Award - IFP.org
Win a Golden Globe - HFPA.org
Win an English Film Award - BAFTA.com
Win an Independent UK Award - BIFA.org.uk
Sundance Film Festival Award - Sundance.org
Cannes Film Festival Award - Festival-Cannes.fr
Europe Film Awards - EuropeanFilmAcademy.org

FILM WORKSHOPS & EDUCATION

Best 10 Film Education Sites

Filmmaking Courses, Directing Workshops, Lighting & Editing Seminars & Regional Independent Film Ed

LA Filmmaking Workshops - IFP.org
NY Film Seminars - FVA.org
Chicago Film Education - ChicagoFilmmakers.org
Seattle Film Workshops - 911Media.org
Atlanta Filmmaking Workshops - ImageFV.org
San Francisco Film Courses - FilmArts.org
Detroit Film Workshops - DetroitFilm.org
USA Film Media Associations - Namac.org
American Film Institute Workshops - AFI.com/Education
Film Director Training - AssistantDirector.com

ACTOR CRAFT & BUSINESS
10 Best Film Acting Sites

Movie Acting techniques, Film Actor Business & Acting Craft: Movie Actor Monologues

Acting Craft - ActingAtoZ.com
Screen Actor Craft - ActorsCraft.com
Acting on the Web - ActingOnTheWeb.com
Acting Business - ActingBiz.com
Actors Resources - ActorSource.com
Actors Chatroom - ActorSite.com
Acting Resources - LAactorsOnline.com
Actors Career Guidance - StarvingActor.com
Film Actor Monologues - ActorPoint.com
Film Acting Monologues - ActorTips.com

ACTOR CASTING
Top 10 Casting Actors Sites

Cast your movie. Hire actors. Casting Calls How to find and cast a film actor or movie star.

Post a Casting Notice - Backstage.com/casting
Actors Directory - HCDonline.com
Movie Actor Resumes - ActingResume.com
Casting Film Actors - Casting-America.com
VIP Film Casting - VIPcasting.com
Cast an Actor - SubmitLink.com
Feature Film Casting - BreakdownServices.com
Daily Casting Calls - CastingCallOnline.com
Find an Actor, Hire a Star - Casting network.com

DIRECTOR NEWS, GOSSIP & UNIONS
Ten Best Director Sites

Want to be a Film Director? Want to hire a movie Director? Learn how to Direct? What to pay a Director?

American Film Directors Union - DGA.org
English Movie Directors Union - DGGB.co.uk
Australian Film Directors Union - asdaFilm.org.au
Canadian Movie Directors Union - DGC.ca
New Zealand Directors Association - sdgNZ.co.nz
Film Director Training - DGAtrainingProgram.org
Assistant Director Training - AssistantDirectors.com
Directors Cast Movie Actors - CastingSociety.com
Film Director Directory - IMDBpro.com
Director Casting Notice - CastList.com

SCREENWRITING RESOURCES
Best 10 Screenwriting Sites

Write a Script! Hire a Screenwriter! Get the Independent Screenwriting facts! Plus much more

Screenwriting A-Z - CreativeScreenwriting.com
Screenwriter Nuts & Bolts - ScreenwritersUtopia.com
Screenwriting Software - WritersStore.com
Selling Scripts & Screenplays - ScriptMag.com
Screenwriting Basics - SydField.com
Screenplay & Teleplay Resources - ScreenStyle.com
Screenwriter Resources - ShowBizSolutions.com

UK Screenwriting Resources - ScreenwriterStore.co.uk
Canadian Writing resources - PraxisFilm.com
Free Script Screenplay Software - Sophocles.net

SCREENWRITER BUSINESS
10 Best Screenwriter Sites

Register a Script! Sell a Screenplay! Find a literary agent! Pitch a story! Sell a Story! Protect an Idea!

Register a Screenplay - WGA.org
Copyright a Script - LOC.gov
Sell a Script - HollywoodLitSales.com
Screenplay Pitches - ScriptShark.com
Script Sales - ScriptPimp.com
Business of Screenwriting - ScreenTalk.Biz
Screenwriter Mentors - WritersChannel.net
Selling Ideas to Hollywood - MoviePitch.com
What is Hollywood Buying - ScriptSales.com
Screenwriting 101 - ScreenwritersUtopia.com

SCREENWRITING COMPETITIONS
Top 10 Script Competition Sites

Win Awards! Screenplay Winners! Secure a Literary Agent! Sell a Script! Screenwriter Awards!

Nicholl Fellowships Award - FellowshipProgram.nick.com
Walt Disney Screenwriting - abcTalent.Disney.com
Sundance Screenwriter Lab - Sundance.org
Be a Director Award - ProjectGreenlight.com
Warner Brothers Award - WarnerBros.com/WritersWorkshop
Chesterfield Script Award - Chesterfield-co.com
Script Competition - WriteMovies.com
Slamdance Script Competition - Slamdance.com/Screencomp
Final Draft Writers Award - FinalDraft.com/BigBreak

SCREENWRITING CONSULTANTS
Ten Best Script Analyst Sites

Planning the screenplay re-write! Hiring a Consultant! Doing a Script re-write! Want a Script Analyzed?

Syd Field - SydField.com
David Trottier - DaveTrottier.com
Dara Marks - DaraMarks.com
MOW Consultant - ScreenplayServices.com
TV Script Consultant - HollywoodScript.com
Natalie Rothenberg - InsiderSystem.com
Melody Jackson - SmartGirlProd.com
Script Coverage - CoverScript.com
Screenplay Analysis - ScriptServices.com
Screenplay Coverage - DeepFeedback.com

A-LIST SCREENWRITERS
10 Famous Screenwriters

Oscar Winners! Box Office Successes!

William Goldman - IMDB.com/name/nm0001279/
Frank Pierson - IMDB.com/name/nm0682757/
Ron Bass - IMDB.com/name/nm0060103/
Robert Towne - IMDB.com/name/nm0001801/

Paul Schrader - IMDB.com/name/nm0001707/
Frank Darabont - IMDB.com/name/nm0001104/
Ruth Jhabvala - IMDB.com/name/nm0695609/
Cameron Crowe - IMDB.com/name/nm0001081/
Tom Stoppard - IMDB.com/name/nm0001779/
Nora Ephron - IMDB.com/name/nm0001188/
Frank Pierson - IMDB.com/name/nm0682757/

FAMOUS INDEPENDENT FILMMAKERS
Top 10 Filmmakers

Screenwriter - Filmmaker! Successful Filmmakers who are also Producers, Directors and Executive Producers

John Cassavetes - People.bu.edu/rcarney/cassavetes
Quentin Tarantino - Tarantino.info
Spike Lee - IMDB.com/name/nm0000490/
Kevin Smith - IMDB.com/name/nm0003620/
John Sayles - IMDB.com/name/nm0000626/
Atom Egoyan - IMDB.com/name/nm0000382/
Woody Allen - IMDB.com/name/nm0000095/
Spike Jonze - IMDB.com/name/nm0005069/
Joel & Ethan Coen - IMDB.com/name/nm0001054/
Neil Labute - IMDB.com/name/nm0001438/

A-LIST HOLLYWOOD DIRECTORS
Top 10 Directors

Blockbuster Film Directors! American Film Directors! Box Office Winners!

Steven Spielberg - IMDB.com/name/nm0000229/
Martin Scorsese - IMDB.com/name/nm0000217/
Frances Coppola - IMDB.com/name/nm0000338/
Michael Mann - IMDB.com/name/nm0000520/
Peter Jackson - IMDB.com/name/nm0001392/
Ron Howard - IMDB.com/name/nm0000165/
Clint Eastwood - IMDB.com/name/nm0000142/
Barry Levinson - IMDB.com/name/nm0001469/
Sydney Pollack - IMDB.com/name/nm0001628/

A-LIST MOVIE PRODUCERS
Ten Best Producers

Mega Box Office Winners! Big Budget Movies! Studio Dealmakers! Hollywood Moguls!

Jerry Bruckheimer - IMDB.com/name/nm0000988/
Joe Roth - IMDB.com/name/nm0005387/
George Lucas - IMDB.com/name/nm0000184/
Brian Grazer - IMDB.com/name/nm0004976/
Norman Lear - IMDB.com/name/nm0005131/
Saul Zaentz - IMDB.com/name/nm0951763/
Dino de Laurentiis - IMDB.com/name/nm0209569/
Howard Kazanjian - IMDB.com/name/nm0443599/
Albert Broccoli - IMDB.com/name/nm0110482/

FILMMAKER INDUSTRY CONTACTS
Best 10 Film Contact Sites

Find a screenwriter, a producer, a director, an actor! Hire a film crew, a cinematographer, a composer, an editor!

Hire a TV Screenwriter - WGA.org
Hire a Film Director - DGA.org
Hire a Movie Actor - SAG.org
Hire a Cinematographer - IATSE-intl.org
Hire a Producer - IMDB.com
Hire Talent & Stars - HCDonline.com
Hire a Film Crew - FilmStaff.com
Hire a Movie Crew - ProductionHUB.com
Hire an English Actor - ActorsEquity.org
Hire a Canadian Actor - Actra.ca

MOVIEMAKING RESOURCES
10 Top Film Resource Sites

Film Computer Programs! Moviemaking Software!
Pre-Production paperwork, forms, budgets and contracts!

Writers Software - WritersStore.com
Writer Supplies (UK) - ScreenwritingStore.com.UK
Filmmaking Contracts - FilmTvContracts.com
Film How-To Books - WebFilmSchool.com
Film & Movie Supplies - StudioDepot.com
Producer Software - Media-Services.com
Filmmaker Books - WebFilmSchool.com/bookstore
Film Budgeting Software - ShowBizSoftware.com
Europe Film Software - ScreenwriterStore.co.uk
Movie Scheduling Programs - ShowBizSoftware.com

BUDGETING & SCHEDULING MOVIES
Top 10 Film Budget Sites

Pre-production! Plan your film shoot. Movie Budgets,
Low-Budget, Shooting Schedules, and Production Boards.

Movie Budget Software - Easy-Budget.com
Filmmaker Budgets - BoilerPlate.net
Independent Film Budgets - JungleSoftware.com
Movie Scheduling - MindStarProds.com
Film Shooting Schedules - EntertainmentPartners.com
Storyboard Software - WritersStore.com
Budget & Schedule Software - ShowBizSoftware.com
Filmmaker Tools - FilmmakerTools.com
Budget Your Movie - FilmBudget.com
Directors Budget Movies - AssistantDirector.com

MOVIE INSURANCE & FILM CONTRACTS
Ten Best Film Law Sites

Movie Equipment Insurance! Filmmaker Completion Bonds!
Film Guarantors! Filmmaking Permits!

Producers Representative - TheMoviePages.com
Film Production Insurance - ProductionInsurance.com
Movie Completion - Bond FFI.com
Film Completion Guarantor - MPBond.com
Film Insurance - ProductionInsurance.com
Entertainment Attorney - MarkLitwak.com
Producers Agent - MediaLawyer.com
Filmmakers Agent - SlossLaw.com
Film Funding Attorneys - VLANY.org
Film & TV Guarantors - ifgBonds.com

FILM PRODUCTION SUPPLIES
Best 10 Film Equipment Sites

35mm & 16mm Film & TV Supplies! Buy 35mm Film! Rent a Movie Camera! Hire a Film Crew! Purchase Movie Supplies!

Kodak Movie Film - Kodak.com/US/en/Motion
Fuji 16MM & 35MM Film - FujiFilm.com
Arriflex 35MM Film Camera - Arriflex.com
Panavision Movie Cameras - Panavision.com
Filmmaking Supplies - StudioDepot.com
Hire a Movie Crew - ProductionHUB.com
Buy Film Equipment - BHphotoVideo.com
Rent Film Equipment - AFCI.org
Hollywood Film Suppliers - LA411.com
New York Film Suppliers - NewYork411.com
London Film Suppliers - TheKnowledheOnline.com

POST-PRODUCTION SUPPLIES
10 Best Post-Production Sites

Find a Film Editor! Select a 35MM Film Laboratory! Get an Answer Print! Film Editors Training Courses

Hollywood 35MM Film Lab - Fotokem.com
Hollywood Union Film Lab - Technicolor.com
New York 16Mm Film Lab - DUart.com
Film Editors Guild - EditorsGuild.com
Movie Editors News - PostMagazine.com
Film Editor Training (West) - VideoSymphony.com
Film Editor Training (East) - FMCtraining.com
California Post-Production - LA411.com
New York Post-Production - NewYork411.com
England Post-Production - TheKnowledgeOnline.com

MOVIE MUSIC & FILM COMPOSER
Top 10 Film Music Sites

Hire a Music Composer! Purchase Music! Music Clearance! Buy & create Sound Effects! ADR, Foley & Looping!

Film Music News - FilmScoreMonthly.com
Movie Music Magazine - FilmMusicMag.com
Music Clearance Information - Clearance.com
Music Clearance Authority - HarryFox.com
ASCAP Music Rights - ASCAP.com
BMI Music Rights - BMI.com
Find Songs for Films - AllMusic.com
Movie Music Clearance - MusicClearance.com
Music Clearance Agency - Clearance.com
Movie Sound Effects - SonicFoundry.com

DIGITAL FILMMAKING NEWS & GOSSIP
Ten Best Digital Filmmaker Sites

Digital News! Digital Cameras! Digital Lighting! Film Transfers! DV to Film! High Def to Film!

DV Magazine - DV.com/magazine
RES Magazine - RES.com
Digital Videography - Videography.com
Digital Videomaker - Videomaker.com
Digital Producer Digital - Producer.com

DV Information - DV.com
Digital Film Information - DVinfo.net
Digital Movie News - DVcentral.org
Digital Film Transfers - DVfilm.com
DV to Film Transfers - eFilm.com

DIGITAL FILM EQUIPMENT & RESOURCES
Best 10 Digital Equipment Sites

Digital Cameras! Digital Equipment! Digital Editing Software!
High Def, High Definition, Digital Video, Mini-DV & HDV!

Sony PD-170, Sony Z-1 - bssc.sel.sony.com/business
Canon XL2 - Consumer.usa.canon.com
DV Mail Order Cameras - BHPhotovideo.com
DV Camera & Gear - DVgear.com
Edit Software Macintosh - apple.com/FinalCutPro
Avid Edit Software - avid.com
Adobe Edit Software - Adobe.com/motion/main.html
Camcorder Reviews - CamCorderInfo.com
DV & Digital Formats - DVformat.com
High Def & HDV Digital - VideoGuys.com

FILM PUBLICITY & MOVIE MARKETING
10 Best Film Publicity Sites

Film Publicist! Film Promotion Sites! Film Festivals!

Film Sales Consultants - FilmFinders.com
Press Agent Society - PRSA.org
Film Publicists - Eppsonline.com
Independent Movie Publicist - DDApr.com
Independent Film Publicist - MediadeLarte.com
Filmmaker News - IndieWire.com
Independent Film Publicity - IndieRep.com
Film Festival Directory - FilmFestivals.com
Hollywood Foreign Press - HFPA.org
Film Publicist - JagPR.com

FILM FESTIVALS
Top 10 Sites

Film Festival Lists! Online Film Festivals!
Film Festival Information!

Film Festival Promo Site - FilmThreat.com
Film Festival Database - FilmFestivalSource.com
Film Festival Directory - FilmFestivals.com
Internet Film Festivals - Cinema-Sites.com
Web Film Festival - ZoieFilms.com
Digital Film Festival - DFilm.com
Film Festival Lists - InsideFilm.com
Film Festival Submissions - WithoutaBox.com
Mini DV Film Festival - HollywoodDVfilmfestival.com
Film Festival Source - FilmFestivalSource.com

USA FILM FESTIVALS
10 Best US Film Festivals

The Best American Film festivals! USA Film Festivals!

Sundance Film Festival - Sundance.org
Slamdance Film Festival - Slamdance.com

Seattle Film Festival - SeattleFilm.com
Los Angeles Film Festival - LAFilmFest.com
Tribeca Film Festival - TribecaFilmFestival.org
Telluride Film Festival - TellurideFilmFestival.com
SXSX Film Festival - sxsx.com
AFI Film Festival - AFIfest.com
Austin Film Festival - AustinFilmFestival.com
Sarasota Film Festival - SarasotaFilmFestival.com

INTERNATIONAL FILM FESTIVALS
Best 10 Foreign Film Festivals

The best foreign film festivals!
Global Film Festivals! Foreign Festivals!

Berlin Film Festival - Berlinale.de
Cannes Film Festival - Festival.Cannes.fr
Toronto Film Festival - e.bell.ca/FilmFest
Raindance Film Festival - Raindance.co.uk
Rotterdam Film Festival - FilmFestivalRotterdam.com
Edinburgh Film Festival - EdfilmFest.org.uk
Tokyo Film Festival - TIFF-jp.net
Korea Pusan Film Festival - PIFF.org
Hong Kong Film Festival - HKIFF.org.hk
Stockholm Film Festival - FilmFestivalen.se

MAJOR THEATRICAL DISTRIBUTORS
10 Best Film Distributor Sites

Movie Studios! The 7 Major Distributors! Theatrical Movie Distributors 1,000 Print Release! Global Distribution!

20th Century Fox Distributor - FoxMovies.com
Disney Movie Studios - Disney.go.com/DisneyPictures
Paramount Movie Distributor - Paramount.com
Universal Film Distributor - UniversalPictures.com
Sony Theatrical Distributor - spe.sony.com/movies
Warner Brothers Distributor - WarnerBros.com
DreamWorks SKG Distributor - Dreamworks.com
Major Movie Distributors - HCDonline.com
English Film Distributors - BFI.org.uk
Australian Film Distribution - AFC.gov.au

INDEPENDENT FILM DISTRIBUTORS
Ten Best Independent Distributors

Mini-Major Distributors! Art Films! Independent Film Distribution! Foreign Film Releases! Cult Distributors

Miramax Movies Distributor - Miramax.com
New Line Films Distributor - NewLine.com
Lions Gate Films - LionsGateFilms.com
Action Film Distribution - YorkEntertainment.com
Association of Global - DistributorsAFMA.com
Low Budget Distributor - CurbEntertainment.com
Foreign Film Distribution - FortissimoFilms.com
Urban & Latino Film Distribution - MaverickEntertainment.cc
Independent Film Distributors - IFTA-online.org
Film & TV Distributor Directory - HCDonline.com

INTERNET MOVIE DISTRIBUTORS
Best 10 Movie Streaming Sites

Distributing on the web! Internet Distribution!
D Film Distribution! Broadband, VOD & Streaming!

Web Distribution - Broadcast.com
Digital Distribution - IFilm.com
Internet Distributing - AtomFilms.com
D Distribution - CinemaNow.com
Theatrical Web Distributing - Entertainndom.com
Streaming Movies & Films - MovieFlix.com
Download Movies & Films - MovieLink.com
Streaming Independent Films - VOD.com
Broadband Movies Sites - BroadbandBuyer.com
Download Cult Films - BijouCafe.com

MOVIE REVIEWS
10 Best Movie Review Sites

Film Reviews! Movie Critics! Film Critiques!
Movie Previews!

Independent Film Reviews - RottenTomatoes.com
Roger Ebert Film Reviews - Suntimes.com/Ebert
Independent Movie Reviews - AintItCoolNews.com
Film Theater Reviews - BoxOff.com
NY Times Film Reviews - NYTimes.com/pages/movies
Entertainment Weekly - Reviews ew.com/ew
Hollywood Movie Reviews - Hollywood.com
Movie Review Directory - MRQE.com
Weekly Movie Reviews - Movies.go.com
Showtimes & Movie Reviews - TVguide.com

FILM JOBS & MOVIE CAREERS
Best Ten Film Job Sites

How to get a movie job! How to launch a film career!
Your first Hollywood job! How to get Film Training!

HBO Film Training - YouthMentoring.com
ABC Television Training - ABCnewTalent.Disney.com
Entertainment Careers - EntertainmentCareers.com
Film Industry Jobs - JobsInTheBiz.com
Movie Industry Jobs - FilmBiz.com/crew
Find a Film Job - CreativePlanet.com
Movie Studio Jobs - FoxCareers.com
Theatrical Distributor Jobs - WBjobs.com
Movie Job Lists - Crew-Net.com
Film Crew Lists - FilmStaff.com

COMMUNITY THEATRES

www.aact.org

How to Start a Theatre Company

The information below is a brief overview of the main issues you need to address when starting a theatre company.
1. Decide on a collaborative structure.
Do you want to run it as a benevolent dictatorship? A cooperative? Or something in-between? How will you choose projects? Who will help with the non-artistic administrative work? Be very clear about the decision-making structure and lines of responsibility. A good place to begin is with the bylaws of other theatre companies. We've included seven of them on this website.

2. Develop a mission statement.
This should be your company's identity, and the litmus test by which you consider future projects. Make it as broad as possible, while still identifying a unique vision. [See The Mission Statement]
3. Create a name and a look.
Brainstorm ideas on names for your company. Develop a logo and an identifiable "look" for your publicity, etc. Check to make sure that your proposed name is not already in local use,

4. Register your company name with the local authorities.
In order to become a registered business and protect the name that you have chosen, you must register with your city. You may have to register as a D.B.A. ("Doing Business As"), effectively registering it as an alias for one or more individuals in the company. In San Francisco, you register with the S.F. Tax Collector's Office. Check to see where this is done in your area. After filing your name, you must announce the creation of your business in a local paper.

5. Open a bank account.
The best way to keep track of your company's financial situation is to open a separate bank account. It also enables you to accept checks written to the company's name. You must be registered with the city in order to open an account under a D.B.A. name. Many banks are offering fee-free everything to business customers, so shop around.

6. Sign up with relevant service organizations.
Join AACT, your local theatre alliance, your state or regional organization. This gives you credibility, and access to lots of ideas and services. Consider joining a local service organization such as Rotary and the Chamber of Commerce.

7. Decide on profit/nonprofit status.
This is one of the most difficult decisions to make, if you choose to go nonprofit, there is bureaucracy and paperwork to navigate. The main advantage is qualifying for grants and special discounts. Foundations can legally give grants only to nonprofjts, and individual donors can write off nonprofit contributions on their taxes. However, most foundations give only to companies with a proven track record. Be realistic about whether the effort is worth it. You can always wait and turn nonprofit in the future. (Fiscal sponsorship is a popular option for small companies, whereby they can apply for grants and access bulk mail rates through a separate nonprofit umbrella organization Often these organizations take a percentage of grant income in return.) Click here for more on incorporating as a nonprofit organization.

8. Make funding decisions.
Will most of your budget depend on grants? The kindness of your family? Dues paid by company members? Earned income such as merchandise sales, auctions, and theme parties? Each has its advantages and drawbacks. Making decisions on funding sources and likely amounts of income will help determine how ambitious your plans should be The Foundation Center's whole purpose is to help non-profits and individuals find money.

9. Strategize potential non-production activities.
Doing a show or two a year is enough for some groups, while others like to maintain a continual presence to increase visibility and develop audiences. Examples of additional activities include teaching classes, hosting staged readings, and engaging in various forms of community outreach.

10. Create a database.
Nothing takes more time than repeatedly writing out addresses on postcards. There are lots of good database programs out there that allow you to print labels in a matter of minutes. It is advisable to keep a separate database for general mailings and another for press contacts.

11. Think About Your Audience.
Who will come see your plays? How are they likely to learn about them? How do similar companies promote their work? Consider doing a list exchange or purchasing mailing lists from companies with audiences you would like to attract.

12. Work Backwards.
Put your first production date far enough into the future that you have time to lay some groundwork, You won't do your best planning under the deadline of an approaching opening night.
13. Join AACT. Membership in the American Association of Community Theatre links you with companies all over the United States and beyond. You'll get the benefits of networking with thousands of other theatre people, and more than 1500 different companies of all sizes, in all kinds of communities. You'll also have access to the members-only pages of this website, including proven ideas for fundraising, choosing a season, publicity and marketing, and much more.

Preparing for your Company's First Show

1. Develop a budget.
Be as specific as possible. Remember any of the following that are applicable: script rights, space rental (liability insurance may be additional), set construction, props, lights rental, costumes, stipends for actors/designers/running crew, various publicity activities, video archiving, and a generous line for "miscellaneous." When in doubt, be conservative about your likely income (especially ticket sales) and be generous in estimating your costs.

2. Secure rights to the play.
If the show is not an original work, be sure you have a signed contract in hand before you announce the play as part of your season. This is generally done in two parts. First, call the playwright/publisher/agent and determine whether the rights are available for your region. If so, then you must make firm arrangements for a performance space and dates, before the playwright/publisher/agent will even consider your request for rights.

3. Contract with a venue.
Space is at a premium in the Bay Area, so be sure to contract many months in advance. Find out what is included in the rental price, such as insurance and tech rehearsals. Many venues require you to acquire your own liability insurance, which can be very expensive. Practically every venue has a different policy on how much rehearsal/tech time is included in the basic rental fee -- be sure to budget the rental cost of any additional tech/dress rehearsals. Consider how long and how often you want perform. The longer a show runs, the more it costs, but also the more people can hear about it and make plans to attend. Weekend matinees vary in their success, based largely on the type of audience members your show attracts. You might consider performing midweek once in a while, so that people working on other shows can see your work.

4. Make union decisions.
If you intend to hire any members of Actors' Equity, you must secure a contract from the Equity office. A BAPP contract may enable you to hire a small number of union actors at very little cost, if you meet certain budget and insurance criteria. A

5. Decide on a system for selling tickets.

Most small companies rely on voice mail for patrons to leave ticket reservations. If you decide to take this route, choose the lucky person who gets to receive those calls. Your local phone company can set up a separate, inexpensive voice mail number, so that nobody's home number is inundated with reservations. Some venues will provide their own box office services. Alternatively, there are services that can handle your ticket sales for a modest fee per ticket. Make sure that you arrange this a couple of months in advance, so that all marketing materials carry the correct box office phone number.

6. Introduce yourself to the press.
The cost of publicity is almost always underestimated. It can sometimes amount to a third of your budget. Especially for your company's first show, it is important to get your name out to the local press through early announcements, a newsletter, a letter of introduction, or whatever other creative method you dream up. Four to eight weeks prior to your opening, send out press releases and photos to all relevant calendars. A professional-looking photo is one of the surest ways to bring extra attention to your calendar listings. Then, a couple of weeks prior to opening, send press packets to local critics, including any background info that might entice them to see the show. AACT's members-only website has a Knowledge Base that includes tips on publicity and marketing, including sample news releases.

California

City	State	Org Name	Office Phone	Box Office Phone	Website
Cambria	CA	Pewter Plough Playhouse	805-927-3877	805-927-3877	http://pewterploughplayhouse.org
Carmel	CA	Children's Experimental Theatre, Inc.	831-624-1531	831-624-1531	http://www.cetstaffplayers.org
Carmichael	CA	Best Of Broadway Benefit Productions, Inc.	916-974-6280	916-974-6280	http://www.bestofbroadway.org
Chino	CA	Chino Community Theatre	909-590-1149	909-590-1149	http://chinocommunitytheatre.org
Chula Vista	CA	OnStage Playhouse	619-422-7787	619-422-7787	http://www.onstageplayhouse.org
Clayton	CA	Rising Moon Marketing and Public Relations	925-672-8717		
Corona	CA	Christian Arts And Theater (CAT)			
Diamond Springs	CA	Imagination Theatre Company	530-642-0404	530-642-0404	http://www.imagination-theater.org
El Cajon	CA	Christian Community Theater	619-588-0206	619-588-0206 or 800-	http://www.cctcyt.org
El Centro	CA	Mostly Theatre Company		760-352-5353	http://www.geocities.com/mostlytheatrecompany
El Cerrito	CA	Contra Costa Civic Theatre	510-524-6654	510-524-9132	http://www.ccct.org
Fontana	CA	Fontana Mummers, Inc.	909-822-4052	909-822-4052	http://www.fontanamummers.com
Fort Irwin	CA	Fort Irwin Music & Dance	760-380-6212		http://www.ftirwinmwr.com
Fresno	CA	Children's Musical Theaterworks		559-442-3140	http://www.cmtworks.net
Fullerton	CA	Muckenthaler Cultural Center	714-738-3135		http://www.muckenthaler.org
Glendale	CA	Stepping Stone Players	818-548-3180		http://steppingstoneplayers.com
Glendora	CA	Renaissance Theatre Company	626-857-4750		
Grass Valley	CA	Community Players of Nevada County, Inc.	530-273-7201	530-273-7201	http://www.communityplayersnc.com

Hayfork	CA	Mountain Actors Workshop, Inc			
Hemet	CA	Ramona Hillside Players	951-926-3243	951-658-5300	http://ramonahillsideplayers.org
Hollister	CA	San Benito Stage Company	831-637-9108		http://www.sanbenitostage.org
La Mesa	CA	Lamplighters Community Theatre	619-464-4598	619-464-4598	http://www.lamplighterstheatre.org
Laguna Beach	CA	No Square Theatre	949-497-1950	949-497-1950	http:// www.nosquare.org
Lake Balboa	CA	Haven Theatre Company			http://www.thehaventheatre.com
Lake Forest	CA	Lake Forest II Show Boaters	714-835-3020 x121		http://www.showboaters.org
Lincoln	CA	Lincoln Hills Players			
Lodi	CA	Gnosis Theatre Company		209-469-3059	http://www.gnosistheatrecompany.org
Lompoc	CA	Lompoc Civic Theatre	805-735-2281		
Los Angeles	CA	Kentwood Players			http://www.kentwoodplayers.org.
Los Angeles	CA	Lyric Opera of Los Angeles	800-471-9515		
Mammoth Lakes	CA	Sierra Classic Theatre, Inc.	760-934-7575		http://sierraclassictheatre.org
Merced	CA	Merced Shakespearefest	209-723-3265		http://www.mercedshakespearetest.org
Merced	CA	Playhouse Merced	209-725-8587	209-725-8587	
Newport Beach	CA	Newport Theatre Arts Center	949-631-0288	949-631-0288	http://ntac-online.com
Old Town Newhall	CA	Canyon Theatre Guild	661-799-2700	661-799-2702	http://www.canyontheatre.org
Oxnard	CA	Teatro De Las Americas	805-485-5445 x2	805-485-5445 x2	
Pacific Palisades	CA	Theatre Palisades, Inc.	310-470-0491	310-454-1970	http://THEATREPALISADES.com
Palmdale	CA	Palmdale Repertory Theatre	213-231-2215 x16		http://PRtheatre.org
Paradise	CA	Theatre On The Ridge	530-872-1768	530-877-5760	http://www.totr.org
Poway	CA	Poway Performing Arts Company		858-679-8085	http://www.powpac.tripod.com
Rancho Cucamonga	CA	Rancho Cucamonga Community Theatre Program	909-477-2760 x2101		
Red Bluff	CA	BareStage Theatre	530-529-1241	530-529-1241	http://www.barestage.com
Redding	CA	Riverfront Playhouse	530-365-1112	530-244-4278	http://www.riverfrontplayhouse.net
Ridgecrest	CA	Community Light Opera & Theatre Association	760-446-2411		http://www.ridgecrest-arts.org
Roseville	CA	Magic Circle Repertory Theatre	916-782-1777	916-782-1777	http://www.mcircle.org
Salinas	CA	ARIEL Theatrical, Inc.	831-775-0976		http://www.arieltheatrical.org
San Bernardino	CA	Junior University of San Bernardino, Inc.	909-887-8118		http://www.junioruniversity.org
San Diego	CA	Scripps Ranch Theatre	858-693-8382	858-578-7728	http://www.scrippsranchtheatre.org
San Jose	CA	Tabard Theatre Company	408-979-0231	408-979-0231	http://www.tabardtheatre.org
San Juan	CA	South Orange County Community	949-248-0808	949-489-8082	http://caminorealplayhouse.org

Capistrano		Theatre			
San Luis Obispo	CA	San Luis Obispo Little Theatre	805-781-3889	805-786-2440	http://www.slolittletheatre.org
Santa Monica	CA	Santa Monica Theatre Guild	310-892-0968	310-828-7519	http://morgan-wixson.org
Solvang	CA	Valley Community Theatre	805-688-3343		
Stockton	CA	Stockton Civic Theatre	209 473-2400	209-473-2424	http://www.sctlivetheatre.org
Sunnyvale	CA	Actors Theatre Center	408-985-5500	408-985-5500	http://www.theatrecenter.biz
Tehachapi	CA	Tehachapi Community Theatre			http://www.tehachapiarts.com
Thousand Oaks	CA	Conejo Players, Inc.	805-495-3715	805-495-3715	http://www.conejoplayers.org
Tustin	CA	Master's Repertory Company	714-393-1079		http://www.mastersrep.org
Twentynine Palms	CA	Theatre 29			http://theatre29.com
Visalia	CA	Enchanted Playhouse Theatre Company		559-739-4600	http://www.enchantedplayhouse.org
Visalia	CA	Visalia Community Players	559-734-3375	559-734-3900	
Westminster	CA	Westminster Community Theatre		714-893-8626	http://www.wctstage.org
Willits	CA	Willits Community Theatre	707-459-0895	707-459-2281	http://www.ALLAboutWCT.org
Willows	CA	Glenn County Theatre Company	530-934-8050		

Georgia

City	State	Org Name	Office Phone	Box Office Phone	Website
Albany	GA	Theatre Albany	229-439-7193	229-439-7141	http://www.theatrealbany.com
Americus	GA	Sumter Players, Inc.	229-924-2645	229-931-0001	http://www.sumterplayer.org
Augusta	GA	Paine College	706-821-8263		http://www.paine.edu
Blue Ridge	GA	Blue Ridge Community Theater	706-632-9223	706-632-9223	http://www.gatewaymountainplayers.com
Carrollton	GA	Encore Theatre Company	770-834-8294	770-830-6106	http://www.encoretheatre.com
Cartersville	GA	Pumphouse Players, Inc.	770-387-2610	770-387-2610	http://www.pumphouseplayers.com
Clarkesville	GA	Habersham Community Theatre	706-754-5654	706-754-5654	http://www.habershamtheater.org
Fort Gordon	GA	Fort Gordon Performing Arts Center	706-791-4389	706-793-8552	http://www.fortgordon.com/fgl.htm
Fort McPherson	GA	IMA Ð South East	404-669-6805		
Macon	GA	Macon Little Theatre	478-477-3342	912-471-PLAY	http://www.maconlittletheatre.org

Martinez	GA	J. S. H. Performing Arts Center	706-868-3349		
McDonough	GA	Henry Players, Inc.	404-765-9000		http://www.henryplayers.com
Newnan	GA	Newnan Theatre Company	770-252-1956	770-683-6282	http://www.newnantheatre.com
Ranger	GA	Tater Patch Players			http://taterpatchplayers.org
Statesboro	GA	Statesboro Theater Arts Regional Alliance, Inc.	912-489-3237		
Stone Mountain	GA	Latino Atlanta Street Theatre			http://www.l-ast.com
Valdosta	GA	Georgia Theatre Conference	229-244-3725		http://www.georgiatheatreconference.com
Valdosta	GA	Little Actors' Theatre	229-249-0822		http://www.littleactors.org

Florida

City	State	Org Name	Office Phone	Box Office Phone	Website
Atlantic Beach	FL	Atlantic Beach Experimental Theatre, Inc.	904-249-7177	904-249-7177	
Bradenton	FL	Manatee Players	941-748-0111	941-748-5875	http://www.manateeplayers.com
Cape Coral	FL	Cultural Park Theatre Company	239-772-5862	239-772-0280	http://culturalpark.com
Clearwater	FL	Clearwater Little Theatre	727-447-5347	727-446-1360	
Clearwater	FL	West Coast Players	727-725-5292	727-734-7100	http://www.wcplayers.com
Dade City	FL	Arts in Motion Community Youth Theatre/Arts Education,			http://www.artsinmotionpasco.org
Daytona Beach	FL	Daytona Playhouse			
Deland	FL	Sands Theater Center, Inc.	386-738-7156	386-736-7456	http://www.sandstheatercenter.com
Eustis	FL	Bay Street Players, Inc.	352-357-7777	352-357-7777	http://www.baystreetplayers.org
Fernandina Beach	FL	Amelia Community Theatre	904-261-6749	904-261-6749	http://AmeliaCommunityTheatre.org
Fernandina Beach	FL	Fernandina Little Theatre	904-321-1595	904-277-2202	
Fort Myers	FL	Out Of The Box Productions	239-357-3956		http://www.getoutofthebox.org
Fort Walton Beach	FL	Stage Crafters Community Theatre	850-243-1101	850-650-6010	http://www.stagecrafters.com
Gainesville	FL	Gainesville Community Playhouse			http://www.gcplayhouse.org
Gulfport	FL	Island Community Theatre, Inc	727-345-1474		http://www.islandcommunitytheare.com
Gulfport	FL	Suncoast Theatre Alliance	727-343-6549		
Hernando	FL	Citrus County Art League, Inc.	352-746-7606	352-746-7606	http://www.artleague.info

Jacksonville	FL	Theatre Jacksonville	904-396-4425	904-396-4425	http://www.theatrejax.com
Jacksonville Beach	FL	Players By The Sea	904-249-0289	904-249-2022	http://www.playersbythesea.org
Lake Wales	FL	Lake Wales Little Theatre	863-676-1266	863-676-tkts (8587)	http://www.lwlt.org
Lake Worth	FL	Lake Worth Playhouse	561-586-6169 x204	561-586-6169	http://www.lakeworthplayhouse.org
Lakeland	FL	Lakeland Community Theatre, Inc. / Pied Piper Players	863-603-7529	863-603-7529	http://www.piedpiperplayers.com
Marathon	FL	Marathon Community Theatre	305-743-0408	305-743-0994	http://MarathonTheater.org
Marco Island	FL	Marco Players			
Miami Beach	FL	Fourth Dimension of South Beach, Inc.	305-532-3740		http://4dsb.org TheaterMania.com
Miami Lakes	FL	Community Theatre of Miami Lakes-Main Street Players	954-675-6421		http://www.mainstreetplayers.com
Naples	FL	Naples Players, Inc.	239-434-7340	239-263-7990	http://www.naplesplayers.org
New Smyrna Beach	FL	Little Theatre of New Smyrna Beach, Inc.	386-423-1246	386-423-1246	http://www.littletheatreofnsb.org
Ocala	FL	Ocala Civic Theatre	352-236-2851	352-236-2274	http://www.ocalacivictheatre.com
Okeechobee	FL	Okeechobee Community Theatre	863-763-1307	863-763-1307	
Pembroke Pines	FL	Pembroke Pines Theatre of the Performing Arts	305-552-4287	877-4PP-TOPA (477-86	http://www.pptopa.com
Pensacola	FL	Pensacola Little Theatre, Inc.	850-434-0257	850-432-2042	http://www.pensacolalittletheatre.com
Pinellas Park	FL	Florida Theatre Conference, Inc.	727-568-0956		http://www.flatheatre.org
Sebring	FL	Highlands Little Theatre, Inc.	863-699-2902	863-382-2525	http://highlandslittletheatre.org
Spring Hill	FL	Stage West, Inc.	352-683-5113	352-683-5113	http://www.stagewest.net
St Petersburg	FL	St. Petersburg Little Theatre, Inc.	727-551-5229	727-866-1973	http://www.splt.info
Stuart	FL	Barn Theatre, Inc.	772-287-4834	772-287-4884	http://www.barn-theatre.com
Tallahassee	FL	Tallahassee Little Theatre	850-224-4597	850-224-8474	http://www.tallahasseelittletheatre.org
Tallahassee	FL	Theatre A La Carte		850-224-8474	http://www.theatrealacarte.org
Titusville	FL	Titusville Playhouse, Inc.	321-268-3711	321-268-1125	http://www.titusvilleplayhouse.net
Venice	FL	Venice Little Theatre	941-484-4033	941-488-1115	http://www.venicestage.com
Vero Beach	FL	Vero Beach Theatre Guild, Inc.	772- 562- 8300	772- 562- 8300	http://www.verobeachtheatreguild.com
Winter Haven	FL	Theatre Winter Haven	863-299-2672	863-294-SHOW(7465)	http://www.theatrewinterhaven.com

Illinois

City	State	Org Name	Office Phone	Box Office Phone	Website
Alton	IL	Alton Little Theater, Inc.	618-462-3205	618-462-	http://www.altonlittletheater.org

				6562	
Belvidere	IL	Main Street Players of Boone County	815-962-5591 x-141	815-229-4444	http://www.mainstreetplayersofboonecounty.com
Bloomington	IL	Community Players	309-663-2121	309-663-2121	http://www.communityplayers.org
Byron	IL	Byron Civic Theatre	815-978-7297	815-234-3000	http://www.bctmagic.com
Champaign	IL	Champaign-Urbana Theatre Company	217-344-3884	217-356-9063	http://www.cutc.org
Channahon	IL	Chanooka Center Stage Theatre Company	815-467-7275		http://www.angelfire.com/musicals/dramas_and_comedy
Chicago	IL	Illinois Theatre Association	773-929-7288 x218		http://www.iltheassoc.org
Danville	IL	Danville Light Opera		217-431-1660	http://www.danvillelightopera.org
Downers Grove	IL	Grove Players	630-415-3682		http://www.groveplayers.org
Effingham	IL	Fine Arts, County of Effingham			http://www.effinghamface.com
Elmhurst	IL	GreenMan Theatre Troupe	630-479-0309	708-447-3514	
Geneva	IL	Rising Star Theatreworks	630-262-5027	630-262-5027	http://www.risingstartheatre.org
LaSalle	IL	Stage 212, Inc.		815-224-3025	http://www.stage212.org
Maryville	IL	Curtain's Up Theater Company		618-692-6994	http://www.curtainsuptheater.com
Mundelein	IL	Kirk Players	847-566-6391	847-566-6391	http://www.kirkplayers.com
Palatine	IL	Theatre Nebula	847-359-9378	847-359-9378	http://theatrenebula.org
Pecatonica	IL	Pec Playhouse Theatre		815-239-1210	http://www.pecplayhouse.org
Peoria	IL	Corn Stock Theatre	309-676-2196	309-696-2196	http://www.cornstocktheatre.com
Peoria	IL	Peoria Players Theatre	309-688-4473	309-688-4473	http://www.PeoriaPlayers.org
Quincy	IL	Quincy Community Theatre	217-222-3209	217-222-3209	http://www.1qct.org
River Forest	IL	AEJ Design, Co	708-655-4866		http://www.aejdesign.com
Romeoville	IL	Association of Region Three (ART)	630-759-6340		
Schaumburg	IL	Schaumburg Park District	847-490-7015		
Waterloo	IL	MASC - Monroe Actors Stage Company	618-337-5876	618-939-7469	http://masctheatre.org

West Chicago	IL	Countryside Players	630-365-5810	630-365-5810	
Western Springs	IL	Theatre of Western Springs	708-246-4043	708-246-3380	http://www.theatrewesternsprings.com
Wilmette	IL	Wilmette Park District	847-256-9694	847-256-9787	

North Carolina

City	State	Org Name	Office Phone	Box Office Phone	Website
Asheville	NC	Asheville Community Theatre	828-254-1320 x7	828-254-1320 x1	http://www.ashevilletheatre.org
Charlotte	NC	Metrolina Theatre Association (MTA)	704-347-1975		http://www.metrolinatheatre.org
Charlotte	NC	Theatre Charlotte	704-376-3777	704-334-9128	http://www.theatrecharlotte.org
Elizabeth City	NC	Encore Theatre Company	252-330-5534	252-338-3382	http://www.encoretheatre.org
Fayetteville	NC	Gilbert Theater	910-485-7012		http://www.theartscouncil.com/gilbert
Goldsboro	NC	StageStruck: The Young People's Own Theatre	919-736-4530		http://www.stagestruck.org
Greensboro	NC	City Arts Drama Center	336-335-6426		http://www.livestockplayers.com
Greensboro	NC	Southeastern Theatre Conference (SETC)	336-272-3645		http://www.setc.org
Hickory	NC	Hickory Community Theatre	828-327-3855	828--328-2283	http://www.hct.org
Kill Devil Hills	NC	Theatre of Dare	252-261-4064		http://www.theatreofdare.org
Littleton	NC	Lakeland Cultural Arts Center, Inc.	252-586-3124	252-586-5577	http://members.tripod.com/`lakelandartscenter/index-.com
Raleigh	NC	North Carolina Theatre Conference	919-832-9171		http://www.nctc.org
Raleigh	NC	Raleigh Little Theatre, Inc.	919-821-4579 x234	919-821-3111	http://www.raleighlittletheatre.org
Rocky Mount	NC	Playhouse Theatre of Rocky Mount	252-972-1441	252-972-1266	http://ci.rocky-mount.nc.us/artscenter/
Sparta	NC	Alleghany Community Theater			http://sparta-nc.com/ACT/
Waynesville	NC	Haywood Arts Regional Theatre	828-456-7956	828-456-6322	http://www.harttheatre.com
Wilson	NC	Playhouse of Wilson	252-291-8951	252-291-8951	http://www.playhouseofwilson.com
Winston Salem	NC	Little Theatre of Winston-Salem	336-748-0857 x202	336-725-4001	http://www.littletheatreonline.com

New York

City	State	Org Name	Office Phone	Box Office Phone	Website
Averill Park	NY	Circle Theatre Players, Inc.	518-674-2007	518-674-2007	http://www.slca-ctp.org
Buffalo	NY	In The Light Theater Company	347-623-6494		
East Aurora	NY	Aurora Players, Inc.	716-622-3113 x118	716-655-1040	http://auroraplayers.org
Elmira	NY	Elmira Little Theatre		607-734-8191	
Garrison	NY	Philipstown Depot Theatre	845-424-4618		
Hadley	NY	Actors Conservatory Theatre	914-391-6558	914-391-6558	http://actshows.org
Hammondsport	NY	Keuka Lake Players, Inc.			http://www.keukalakeplayers.org
Harrison	NY	Westchester Group Theatre, Inc.	914-509-5204		
Honeoye Falls	NY	Greater Rochester Repertory Companies, Inc.	585-624-4608	585-624-8360 or 5856	http://www.grrc-arts.org
Huntington	NY	Township Theatre Group	631-421-9832		
Monroe	NY	Creative Theatre Group, Inc.	973-357-3282		http://www.ct-mwp.org
Naples	NY	Theatre Association of New York State (TANYS)	315-789-1209		http://www.tanys.org
New Hartford	NY	Players of Utica		315-724-4000	http://www.playersofutica.org
North Creek	NY	Our Town Theatre Group		518-251-2938	http://www.ourtowntheatergroup.org
Nyack	NY	Elmwood Community Playhouse, Inc.	914-536-6624	845-353-1313	
Olean	NY	Olean Community Theatre, Inc.			http://www.OCTNOW.com
Owego	NY	Ti-Ahwaga Community Players, Inc.	607-687-2130	607-687-2130	http://www.tiahwaga.com
Peru	NY	Essex Theatre Company, Inc.	518-643-0888	518-643-0888	
Point Lookout	NY	Long Beach Theatre Guild, Inc.	516-432-2600	516-432-2600	http://www.longbeachtheatreguild.clickhere2.net
Putman Valley	NY	Spotlight Theatre Productions, Inc.	845-526-3461	845-526-3461	http://www.stpny.com
Rochester	NY	Irondequoit Theatre Guild		585-234-4349	
Saranac Lake	NY	Community Theatre Players, Inc.	518-897-2209	518-891-7362	
Sherburne	NY	Sherburne Music Theater Society	607-674-2561		
Staten Island	NY	Staten Island Shakespearean Theatre Co.		718-390-6616	http://www.lunacat.net/sistcompany/main.htm
Syracuse	NY	USITT - US Institute for Theatre Technology	800-938-7488 or 315-		http://www.usitt.org
Wappingers Falls	NY	County Players, Inc.	845-298-1491	845-298-1491	http://www.countyplayers.org

South Carolina

City	State	Org Name	Office Phone	Box Office Phone	Website
Aiken	SC	Aiken Community Theatre			http://www.atlantic.net/~acp
Aiken	SC	City of Aiken	803-642-7631		
Anderson	SC	Electric City Playhouse	864-224-4248	864-224-4248	http://ecplayhouse.com
Charleston	SC	Footlight Players, Inc.	843-722-7521	803-722-4487	http://footlightplayers.net
Clemson	SC	Clemson Little Theatre	864-646-8100	864-646-8100	http://www.clemsonlittletheatre.com
Clover	SC	Clover Community Theatre	704-355-8202		http://www.geocites.com/clovercommunitytheatre
Fort Jackson	SC	Perez Fitness Center	803-751-7146		
Georgetown	SC	Swamp Fox Players	843-527-2924		
Laurens	SC	Laurens County Community Theatre	864-984-8325	864-833-5228	http://www.lcct.net
Rock Hill	SC	Rock Hill Community Theatre		803-324-2578	http://www.rhcommunitytheatre.org
Rock Hill	SC	TheatreUSA.com	803-325-8428		http://www.theatreusa.com
Seneca	SC	Oconee Community Theatre	864-882-1910	864-882-7700	http://www.oconeetheater.org
Sumter	SC	Sumter Little Theatre	803-775-2150	803-775-2150	http://www.sumterlittletheatre.org
York	SC	Yorkville Players	877-224-6905		http://www.yorkvilleplayers.com

Community Theatre Websites

www.nicoth.com
http://theatreontario.org/content/on_stage.htm
www.lcc.hawaii.edu/org/th/
www.pueblo.org/community/perf_arts.html
www.baycolonyproductions.com/index2.ivnu
http://www.nationaltheatre.org.uk/?lid=2371
www.londontheatre.co.uk/
http://www.playbill.com/index.php
http://www.royalcourttheatre.com
http://www.theatre-resources.net/
http://www.actorsite.com/
http://www.theatredirectories.com/
http://www.artslynx.org/
http://www.caea.com/EquityWeb/default.aspx
http://www.castingnetwork.net/home/default.aspx
http://www.perspicacity.com/
http://www.perspicacity.com/
http://www.unc.edu/depts/outdoor/
http://lecatr.people.wm.edu/
http://www.setc.org/
http://www.bridgesweb.com/blacktheatre/blackcomp.html
http://www.blacktheatrenetwork.org/
http://www.georgiatheatreconference.com/community.htm
http://www.cnyarts.com
http://newyork.citysearch.com/find/section/newyork/events.html
http://www.offbroadwayonline.com/aboutartny.php
http://www.tanys.homestead.com/
http://california.local.com
http://www.californiamusicaltheatre.com/
http://www.at-la.com/@la-perf.htm
http://www.theatrela.org/
http://www.nohoartsdistrict.com/
http://www.sciway.net/tourism/shows-columbia.html
http://www.theatreusa.com/dcd/Promote_in_Theatres/North_Carolina/
http://www.curtainrising.com/usa/usa_nc.html
http://dmoz.org/Arts/Performing_Arts/Theatre ... Theatre/United_States/Florida/
http://www.floridasmart.com/artentertain/theaters.htm

TURBO CHARGE YOUR CAREER!
WRITE AND PRODUCE YOUR OWN STAGE PLAY!

www.ImageBuildersEntertainment.com.

By Beverly Aarons

Have you ever wondered how you can adapt your screenplay or novel into a stage play? Do you have a story that you yearn to see on stage? Or, maybe you're an actor frustrated by the barrage of fruitless auditions and dream of performing before a live audience. If so, writing and producing your own stage play just may be the ticket to turbo charging your career.

Despite urban mythology, Theatre is an art form for the masses, not some stuffy reprieve for the elite. It can be produced by anyone, anywhere and delivered directly to the audience without the meddling of middlemen. Anyone can produce theatre, I did it and so can you.

Taking The Plunge

Are you ready to step out of your rut? There are five easy steps to producing your own theatrical show.

1. Write the script. Keep it small and simple. No fancy sets or a cast of thousands. One or two characters are more than enough to create a compelling story.

2. Find a venue. You don't need to premiere your show at the Kodak Theatre or Pantages. Scope out local community theatres and nearby universities. There are even venues that will allow you to use their space for a nominal fee (as little as $20), as long as you're not selling tickets.

3. Cast and crew up! Find actors and crew amongst your circle of family and friends or check out the college theatre department.

4. Tell the world! Create a press release telling the world about what you're doing. Send it to local newspapers and arts publications. Post online in message groups and arts calendars.

5. Perform your show. You can perform your show once or 20 times. It's up to you. Just remember to continue to market your show throughout the duration of your run.

Don't be afraid to step out there! With a plan and action you can begin producing theatre today. Get more tips on how to produce theatre from the new ebook "How To Write And Produce Your Own Stage Play And Make A Profit" at www.ImageBuildersEntertainment.com.

ABOUT THE AUTHOR
Beverly Aarons is the winner of the Bill Cosby Screenwriting award and the writer/producer of the hit comedic stage play "Confessions of a Creamy Crack

Make Profits From Producing Theatre!
"This book actually explains every step from the beginning to the end...I can't wait to launch my play!" - K. Williams

"Before reading this book, I thought that the process of writing, producing and putting on a play would be really long. But after reading this book, I realized that the process can be a very quick one! ", M. Fuller

"It's easy for the average reader to understand. No extra big words that can seem intimidating. Just straight forward facts!" A. Fisher

Are you a writer, actor or performer who's tired of getting the run around from Hollywood and Broadway?

I once was like you...But in 2003 all that changed.

With only $1500 I premiered my successful stage play in North Hollywood, CA to a packed house.
Find out how I did it, and how you can do it too! Pick up my new book today!

TALENT WEB SITES

www.adamgobble.com

Kidz Entertainment Biz

Industry Links for Children Seeking an Acting Career

www.thetalentconnection.net	www.zedcardprinters.com	www.sag.com	www.reproductions.com
www.marcsphoto.com	www.kyerworkshop.com	www.lacasting.com	www.bbskids.com
Coogan Account Information	www.ChildreninFilm.com	www.minorcon.org	www.imdb.com

http://www2.warnerbros.com/web/main/kids/kids.jsp

www.kidzworld.com
Media and resource websites for child entertainers

www.auditions.net

The site is a casting calls listing site with a collection of acting auditions gathered from throughout the World Wide Web and categorized for your convenience.

www.imdb.com
Welcome to the Internet Movie Database, the biggest, best, most award-winning movie site on the planet.

www.showbizmonster.com

What are the benefits Casting Directors, Talent Agents/Managers Mailing Labels?

Mailing labels are pre-printed labels with the names and addresses of Talent Agents, Casting Directors, and Personal Managers.

Mailing Label Sample

CBS Television Casting Peter Golden 7800 Beverly Blvd. 284 Los Angeles, CA 90036

Save Time
Can you imagine looking up addresses to Casting Directors, Talent Agents, and Personal Managers, then writing them on your flyers or other mailers? Neither can we. No more writing the names and addresses of each contact. When you are ready to do mailers, simply peel off, stick, and mail.

Be Professional
Hand addressed flyers and envelopes look tacky and get tossed in the trash bin. Send mailers to Casting Directors, Talent Agents, and Personal Managers that look professional.

Listings and Directories for the Entertainment Industry

Why is it important to mail to Casting Directors?

Mail your headshot, resume and a cover letter to Casting Directors to get maximum exposure. The function of Casting Directors is to find performers to audition for the directors or producers. Casting Directors organize the casting process and coordinate auditions. Casting Directors use different sources for talent search such as Agents and Managers; however Casting Directors have their own files, which they create from submissions received by mail.

How often should I mail?

To Casting Directors you should mail at least once every 1-2 months. To Agents/Managers you should mail at least twice a month until you sign on with an Agent or Manager.

www.actorsfund.org

The Actors' Fund of America, a nonprofit organization founded in 1882, provides for the social welfare of all entertainment professionals--designers, writers, sound technicians, musicians, dancers, administrators, directors, film editors, and stagehands--as well as actors. Professionals in film, television, radio, theatre, dance and music all turn to The Actors' Fund in times of need. Headquarters and The Aurora Residence are located in New York City, with regional offices in Los Angeles and Chicago. The Lillian Booth Actors' Home of the Actors' Fund of America in Englewood, New Jersey. The Actors' Work Program has offices in New York and Los Angeles.

The Actors' Fund's mission is "to advance, foster and benefit the welfare of all professionals in the entertainment community who are in need of help, ensuring that these efforts are accomplished with compassion, confidentiality and preservation of dignity for the individuals concerned."

www.atlantaperforms.com

Atlanta Coalition of Performing Arts (ACPA) is the nonprofit alliance of 146 performing and associated arts & cultural organizations, along with over 350 member performing arts professionals, from 12 greater Atlanta counties.

Our service organization provides programs & services, which support and promote greater Atlanta's vibrant theatre, dance, music, and film arts. We are pleased to be your guide to our region's performing arts scene with comprehensive information resources like our member listings/links, show calendars and more.

www.auditionforhollywood.com

Gregory Cascante has achieved impressive results during his business career. He is currently **President and Chief Executive Officer of CEO Films LLC**, and **August Entertainment**, a company he founded in 1988. Gregory has been directly involved with more than **40 feature films** during a **27-year career** as either producer or distributor, or both. His senior level management positions have included:

- Founder and President of August Entertainment 1/88 - Present
- Founder and President of CEO FILMS LLC 3/97 Present
- President and CEO of Kushner-Locke International 9/94 - 4/97
- Founder and President of Interaccess Film Distribution (IFD), Vestron's International Subsidiary.
- President and Chief Operating Officer of PRODUCERS SALES ORGANIZATION.
- Vice President in charge of the International Division of TIME-LIFE FILMS.
- Vice President of Acquisitions and Corporate Planning for TIME-LIFE TELEVISION.
- Director of Acquisitions and Programming for HOME BOX OFFICE
- Assistant Circulation Director for TIME magazine.
- Circulation Director of HARVARD BUSINESS REVIEW.

These are some of the many films with which Mr. Cascante has been directly involved either as producer or distributor and to which he has personally contributed: **TRUE ROMANCE, LAWNMOWER MAN, GLENGARRY GLEN ROSS, DIRTY DANCING, NEVER ENDING STORY, NEVER SAY NEVER AGAIN, THE COTTON CLUB, PINOCCHIO, PRIZZI'S HONOR, 9 1/2 WEEKS, FREEWAY, WHOLE WIDE WORLD, THE OMEGA CODE and FOODFIGHT!** (which is currently in production in Los Angeles).

www.blacktalentnews.com

Black Talent News*'* mission is to provide — and act as a network for the exchange of — news, information, and education to both African Americans and people throughout the world to the "Black Entertainment Experience," specifically in the television, film, cable, theatre and music industries.

www.fade11black.com

My name is Hamin "TIGER" Phillips I am the founder of FADE 2 BLACK ENT. I am an Actor/Model, you can put my name in any search engine and read some of my credits. My focus with FADE 2 BLACK ENT is to help talent obtain the marketing tools they need to best represent themselves. When I started in the business five years ago, it took me forever to learn the industry because of the misdirection and the frauds out there. I want to help talent get exactly what they need to obtain an agent and to book their own work all in one trip. We are located in Racine, Wisconsin. We make it possible to work with talent nationwide because of our network of industry professionals.

www.colonize.com/rtv/casting.php

Reality TV Casting Information: The Reality TV database is an independent listing service provided by Colonize. Shows are owned by their respective trademark, servicemark and copyright holders and are not affiliated with Colonize. Colonize does not participate in and has no control over casting decisions.

www.starsearchcasting.com

Star Search Casting was founded in 1998 by Jonathan Parker, an entrepreneur and entertainment industry executive. From a small, pioneering company, Star Search has grown to over 120,000 members and is known as the number one site for stars and aspiring stars of every description to find work in the entertainment business.

Jonathan's goal was to make it possible for anybody, anywhere in the world, even those in remote areas, to have an equal opportunity to be discovered. Jonathan perceived, through the use of the Internet, a shift in the way the industry would do business in the future. This shift would not only allow individuals to be more proactive in seeking out available opportunities, but also allow casting directors and agents to easily find the talent they're searching for from a much larger pool of individuals.

Showcasing talent from Moscow to London, from New York to Los Angeles and every small town in between, Jonathan's vision has been realized. Thousands of new casting notices and auditions each week have eliminated the number one nemeses for the talent--waiting for the phone to ring. Jonathan's philosophy has been to showcase a company of integrity, where its customers enjoy the most comprehensive services and tools available. Star Search Casting has spent years developing the most sophisticated interactive Talent Database on the Internet. This easy-to-use state of the art system allows casting directors to search for talent using dozens of highly specific parameters to enable quick access to the exact person they need. For this reason, thousands of entertainment professionals whose job it is to hire talent, say Star Search Casting is their number one choice among all online casting companies.

Our purpose at Star Search Casting is to provide the highest level of service to both our Talent (members) and those who come to our site to find talent to fill the thousands of new entertainment jobs that become available each week. Our goal is to provide every effective tool we can to enable our Talent to find work in the industry. This is why we provide our Talent Database to casting directors, scouts, agents and others who hire talent, entirely free of charge.

With over 20 million hits per month, and a global reach, Star Search Casting has not only helped to change an industry, but is a company whose leadership and philosophy has helped thousands of stars and aspiring stars live their dreams.

www.onlinetalentdirectory.com

This **FREE** service is available worldwide — just a mouse-click away — from any computer hooked up to the Internet.

It doesn't matter whether you're an established screen actor in Hollywood or a student actor at a small community theater in Moscow, Russia — Are you an actor with your own Web site, or just looking for that **Official Web site** of that favorite actor of yours, this is **your directory**.

This service is **FREE**! It costs nothing to add your site to the directory or look up the URL of any listed Web sites. Please help us grow by **adding your own Web page** or any others that you might know of: Friends that are struggling actors, or that Official Web site of your favorite **Movie Star**.

As long as you're an actor and have your own Official Web site, its URL belongs right here at the **OnlineTalentDirectory.com**.

www.hollywoodnorth.com

The Purpose: Hollywood North has been created as a one-stop solution for seeking, finding, and managing all types of Acting and Modeling careers.

Today, the Hollywood North community continues to grow with more and more Actors, Models and Agents looking to us. Aspiring individuals have a relaxed and safe environment to meet others in the industry from all across the globe.

Best of all, Hollywood North is open 24 hours a day offering world-wide media exposure.

The Difference: Hollywood North offers more than just a static directory of agents looking for Actors and Models. People join us for a sense of community and for the ability to communicate with other people who have similar interests and desires. Without the limitations of geography, our members can develop their portfolios and make connections with members from around the world!

The Evolution: Hollywood North is constantly evolving to meet the wants and needs of its members. Our team is continually exploring and developing new features according to the results of member polls and email feedback. To help members make quality connections, we've created extended profile questions.

The evolution of Hollywood North goes on - with changes, improvements and features constantly in the works to make our site the very best in the entertainment industry!

The Company: Hollywood North is a leader in the global online agency network, allowing registered members to confidentially and anonymously meet in a safe and fun environment. Hollywood North operates with a staff of highly qualified professionals to make your journey to be discovered a lot easier.

www.casting-call.us

The largest Free on-line database
More auditions & castings than all other sites combined

www.craigslist.org

Classifieds throughout the USA, Go to the TV/film/video and talent link

www.kittyboy.com/imagefv/

IMAGE Film & Video
Independent Media Artists of Georgia, Etc.

www.ontopservices.com/icandy360.htm

We take care of all yours needs, get you paid assignments, answer phones, pre-screen clients, and book & confirm appointments, set up all your traveling arrangements, and take care of all your advertising needs.
If you are looking for serious financial gain, and wanting to make at least $500 to $2, 000 a week, depending on your work ethic, ICandy Management is the one for you!

www.love2act.com

A global resource for actors to help establish a successful career

www.exploretalent.com

The World's biggest talent website.

www.alexamodels.com

Alexa Model & Talent Management, Inc.

www.proactorsstudio.com

Acting Classes in Atlanta, Georgia with Acting Instructor Nick Conti

www.kidshollywood.com

The Kids Hollywood Connection was founded twenty five years ago by Phyllis Henson, a mom whose three children appeared in over 150 national commercials. Phyllis has vowed to help others avoid the mistakes she had made. She founded The Kids Hollywood Connection, and has helped at least 10,000 others avoid the non-reputable companies that people hear about in the news almost on a daily basis.

KIDS HOLLYWOOD CONNECTION IS BONDED AND LICENSED AS AN "ADVANCED FEE TALENT SERVICE" THROUGH THE CA STATE LABOR COMMISSIONER'S OFFICE AND FOLLOWS ALL REQUIRED STATE REGULATIONS. KIDS HOLLYWOOD CONNECTION IS NOT A TALENT OR MODELING AGENT AND DOES NOT ACT IN SUCH A CAPACITY. KIDS HOLLYWOOD CONNECTION

ADVISES, CONSULTS AND PLACES CHILDREN WITH SAG FRANCHISED AGENCIES IN THE BUSINESS.

www.hollywoodhandbook.com

The Hollywood Group, Inc. is the brainchild of a group of entertainment pros that have come to recognize that much could be done to make the entertainment business more accessible to the newcomer. A successful entrée into the show business no longer has to be a daunting task. This group of professionals, comprised of Acting Teachers, Photographers, Casting Directors and extra placement services, and Managers and Talent Agents has designed the Hollywood Handbook, a comprehensive program. It is a quality-assured program that offers children and teenagers the preparation tools needed to begin to work in entertainment.

The Hollywood Group's program consists of information, referrals, discounts and ongoing support to you and your child(ren) as part of their preparation for working as actors. It culminates in the industry's most successful showcase where your kid gets to read for professional talent agents and managers. The true measure of a kid's readiness for acting is the willingness of an agent to represent her or him. The true measure of a development company's effectiveness is the average showcase call-back rate its kids experience; ours is a whopping 45%.

www.hollywoodbreakdowns.com

We posted a database of talent on the internet in 1994 and it quickly grew to over 5,000 portfolio listings. Immediately the talent listings received talent inquiries from all across the USA, France, Italy, Japan, Australia etc. The internet was a new media and everyone was excited about it's potential use in casting.

Our experience over the years showed that photos posted on the internet by the talent didn't work well. It's was time consuming for a casting director to look through photo databases, many photos of the talent were not kept current and the response from the talent was delayed if they ever responded at all thus missing casting deadlines.

Sending casting notices directly to actors does not work either. This results in the casting directors being harassed by some actors. Actors looking for professional work should seek representation in Los Angeles and New York. We recommend checking with the Screen Actors Guild. If you are just starting out try

CENTRAL CASTING
Central does extra casting for non-union and union adults for most of the TV Shows in Hollywood. You must register in person, they do not accept photos by mail. They are located at 220 South Flower St. in Burbank, CA

Kids Background
Kids does extra casting for non-union and union children for many of the TV Shows in Hollywood including Disney. You must register in person, they do not accept photos by mail. Kids is located directly across the street from Central Casting on Flower St. in Burbank, CA

www.hollywoodnorthextras.com/cif.htm

At Hollywood North Extras we do require children age 5 and up as movie extras for scenes such as parks, schools, weddings, birthday parties and more. It is normally a fun and exciting opportunity for kids and a chance for parents to see if their kids enjoy being on set, especially if they are considering acting as something they want to pursue. We are not a "children's agency" but supply whatever age is needed. We have worked on many fun projects for kids including Catwoman, Santa Clause 2, Spymate, Elf, Mr. St. Nick and also commercials such as Sony and Kool-Aid.

Some parents consider having both a principal agency, (mostly auditions for commercials), as well as having us as an extras agency. For most principal agents this is not a conflict of interest. Often kids who have a principal agency get the opportunity to audition, but rarely get the opportunity to experience a set. We provide that opportunity, so it gets kids used to the "lights, camera, action" part of it. Of course, for most kids it's just a great way to spend the day, have fun, meet new friends and make some spending money.

Our extras info, filming is usually Monday-Friday so the child will miss a day of school if it is during the school year and will require a parent or chaperone on set with them at all times. Please read our extras info as most of that information applies to booking children as well. As with adults, when we call to book your child and they are unable to miss school that day or nobody can take them, that's okay. We will call again. Please do not say yes unless you can commit for the whole day. We find that most of our children get booked approximately 2-10 times a year. Some of our parents find this an incentive for kids to have good grades, so that when we call to book them it's okay for them to miss a day of school!

The rules have changed for children working on set as of December 2003. The new rules are as per the British Columbia Employment Standards Act for children working in the BC film and television industry. Children no longer need to have a child employment permit card as before but do need to have a 2 page "Written Consent form by Parent or Guardian" filled out prior to working on each production. It is best to download this from the site and have it filled out before arriving on set. It is also available at our office or we can fax it to you from our office. Normally these forms can be obtained and filled out on set prior to working, but it is most convenient if done beforehand. This applies to all children 14 and under. Parents or guardians of children working in film and television should be familiar with the employment standard act from the Government of B.C. Please go to the employment standards WebPages and read their information.

www.childreninfilm.com

Do you have questions relating to employment labor laws of children in film, television, commercials or print photography? ...We've got answers to all kinds of questions, even if you're not sure what questions to ask!

ChildrenInFilm.com is built on over 18 years of experience in supporting Producers, Agents & Parents.

Need to know how to get a child entertainment work permit or information on obtaining a work permit for minors in California? ...We'll show you how!

Want to know more about how to employ your child's creativity?...We offer ten easy steps to success.

- **Some child labor laws in the United States require minors to have a limited workday based on age?**
- **California employment labor law requires some minors to carry a current and valid child entertainment work permit before accepting employment in the Entertainment Industry?**
- **California law requires that parents provide producers with a Coogan Trust document and for producers to make timely payments?**
- **Kids traveling from a distant location to work in Vancouver, British Columbia require five days of private tutoring in every seven?**
- **CA employment law states emancipation does not mean your child can work as an adult?**

www.actingland.com

Looking for casting notices? We've got 'em and we're adding new casting notices all the time. Plus, our casting notices are not limited to just Hollywood or New York City. You'll find casting calls and auditions in areas all around the globe. So browse our list of casting notices to find auditions and interviews in your area.

www.hollywoodauditions.com

HollywoodAuditions.com™ is the fastest-growing online advertising and casting network, helping production companies cast talent since 1996. Our main goal is to bring both ends of the entertainment world together through technology and the Internet. Our website was created to help performers, actors, models, musicians, talent agencies, casting directors, production companies, schools, photographers, and anyone looking to establish contacts in the entertainment industry.

www.thecastlist.com

The only full service on-line casting office, TheCastList.Com is the first casting site to cover all the bases. We are equally useful to industry professionals and actors. By providing a magnificent casting tool and information site to the worldwide entertainment industry, we've positioned ourselves to also be a casting site where new actors have a realistic chance at being seen and cast via the internet.

www.starseachcasting.com

Star Search Casting was founded in 1998 by Jonathan Parker, an entrepreneur and entertainment industry executive. From a small, pioneering company, Star Search has grown to over 120,000 members and is known as the number one site for stars and aspiring stars of every description to find work in the entertainment business.

Jonathan's goal was to make it possible for anybody, anywhere in the world, even those in remote areas, to have an equal opportunity to be discovered. Jonathan perceived, through the use of the Internet, a shift in the way the industry would do business in the future. This shift would not only allow individuals to be more proactive in seeking out available opportunities, but also allow casting directors and agents to easily find the talent they're searching for from a much larger pool of individuals.

www.georgia.org/film

The Georgia Film, Video & Music Office is a state agency housed in the Commissioner's Office of the Georgia Department of Economic Development. Resource information or film office services are offered free of charge to individuals or companies interested in relocating or shooting in Georgia. The Georgia Film, Video & Music Office is a state agency housed in the Commissioner's Office of the Georgia Department of Economic Development. Resource information or film office services are offered free of charge to individuals or companies interested in relocating or shooting in Georgia.

http://www.geocities.com/akatsavou/casting_en.html

Casting & audition links. On-line job opportunities to actors. Many sites that are indexed here are for free services. (The following links are opening in a new window).

- **AACTION** for Canadian talent | **free**, needs registration
- **Actor Point - Casting** (US) | requires payment
- **Actors On Line** Latin America casting notices | **free**
- **Actors equity - Casting Call Search** | **free**
- **Artist website** (US) casting notices by mail | requires payment
- **Audition Finder** | **free**, needs registration
- **Back Stage** (US) casting notices | **free sample**
- **BIGBreakNY** - Entertainment Industry Resource | **free**
- **Casting America** (US) | requires payment
- **Casting Connection** notices, resume posting | requires payment
- **Cast Net** (US) submission service | requires payment
- **Casting you** (US) | requires payment
- **Disney Careers** search various jobs, not only acting | **free**
- **Easy Casting** notices, resume posting | **free**
- **Entertainment Careers** (US), good source | **free**
- **Eperformer** (US) | **free** (but not updated)
- **Free Casting Report** (UK) | **free**

- **Hollywood Reporter Classifieds | free**
- **My8by10 - Free Casting Notices** | **free**
- **New England Casting** talent database | requires payment
- **NWTheatre** Casting notices & resources | **free**
- **Performers directory** (UK) notices | requires payment
- **Performing Jobs** (US) | **free** for performers
- **Playbill Casting & Jobs** On line Casting Notices | **free**
- **ProductionHUB - Classifieds/ Casting Notices** | free
- **Regional, Combined Theatre Auditions | free**
- **Showbiz Monster** posted auditions | **free**
- **Stage Pool™** (Europe) | requires payment
- **Star Maker Casting Notices** | requires payment for listing
- **Talent Bank** make your portfolio | **free**, needs registration
- **Talent calls** posting service (USA) | **free**, needs registration
- **Talent Source** entertainment database | requires payment
- **Texas Auditions-HotBoards | free**
- **The Acting Depot** info for various sections | requires payment
- **TheBuzzNYC** | requires payment
- **The Casting Workbook** (CA) | requires payment
- **The Stage** Magazine's job section on-line (UK) | **free**
- **UKscreen** film crewing & casting | **free**, registration required for more options

MULTICULTURAL AGENCY & MANAGERS LIST

Arts Entertainment [Mngmnt]
9460 Wilshire Blvd. 7th Floor
Beverly Hills, CA 90212
(310) 888-3200/(212) 262-6565

9 Mile Management [Mngmnt]
11271 Ventura Blvd. #329
Studio City, CA 91604
chris@9milemanagement.com
www.9milemanagement.com
Notes: Los Angeles based writers only. E-mail logline, brief bio and synopsis. Will reply if there is interest. No phone calls or unsolicited scripts, please.

Above The Line Agency
9200 Sunset Blvd., #804
West Hollywood, CA 90069
(310) 859-6115

Abrams Artists Agency
9200 Sunset Blvd., 11th Floor
West Hollywood, CA 90069
(310) 859-0625

ACME Talent & Literary Agency
4727 Wilshire Blvd., Suite 333
Los Angeles, CA, 90010
(323) 954-2263

AEI/Atchity Editorial/Entertainment International [Mngmnt]
9601 Wilshire Blvd. Box 1202
Beverly Hills, CA 90210
(323) 932-0407
Email: Webaei@aol.com

Agency For The Performing Arts
9200 Sunset Blvd., #900
Los Angeles, CA 90069
(310) 888-4200
Agency, The
(818) 754-2000

Aimee Entertainment Assoc.
15840 Ventura Blvd., Ste. 215
Encino, CA 91436
(818) 783-3831

Alpern Group, The
15645 Royal Oak Rd.
Encino, CA 91436
(818) 528-1111

Alter Ego Entertainment [Mngmnt]
10474 Santa Monica Boulevard, Suite 403
Los Angeles, CA 90025
United States
310-474-3737

Amsel, Eisenstadt & Frazier, Inc.
5757 Wilshire Blvd., Ste. 510
Los Angeles, CA 90036
(323) 939-1188

Amsterdam Agency, Marcia
41 W. 82nd St. Ste: 9A
New York, NY 10024
(212) 873-4945

AniManagement
12301 Wilshire Blvd, Ste 520
Los Angeles, CA 90025
(310) 207-7100
(310) 207-7111 fax

Anonymous Content [Mngmnt]
8522 National Blvd., Ste. 101
Culver City, CA 90232
(310) 558-3667
(310) 558-4212 fax

Artists Group, Ltd., The
10100 Santa Monica Blvd., #2490
Los Angeles, CA 90067
(310) 552-1100

Artist International Management [Mngmnt]
Attn: Joel Newton
8730 Sunset Blvd., 6th Floor
West Hollywood, CA 90069
(310) 358-9007
Note: Primarily represent TV and Feature writers. Some Directors. Submissions by mail accepted. (One per year per artist)

Bauer Company, The [Mngmnt]
9465 Wilshire Blvd., Ste. 308
Beverly Hills, CA 90212
310-247-3880

Baumgarten Merims Ent. Mgmt [Mngmnt]
1640 S. Sepulveda Blvd., Ste 218
Los Angeles, CA 90025
(310) 996-1885

Becker Entertainment Group [Mngmnt]
6063 Sunset Blvd.
Hollywood, CA 90028
323-461-6055
323-461-6744 fax

Becsey, Wisdom, Kalajian
9200 Sunset Blvd, Ste. 820
Los Angeles, CA 90069
(310) 550-0535

BenderSpink Management (Chris Bender & JC Spink) [Mngmnt]
110 S. Fairfax Avenue, Suite 350
Los Angeles, CA 90036
323-904-1800
323-904-1802 fax

Black Management, Michael [Mngmnt]
5750 Wilshire Blvd., Ste. 640
Los Angeles, CA 90036
(323) 965-2530

Black Talent & Literary Agency, Bonnie
5318 Wilkinson Ave
Valley Village, CA 91607
818-753-5424

Blair Silver & Company Entertainment Management [Mngmnt]
P.O. Box 3188
Manhattan Beach, CA 90266
310-546-4669
blairsilver@aol.com

Blue Train Entertainment [Mngmnt]
9100 Wilshire Boulevard, Suite 340 West
Beverly Hills, CA 90212
310-285-7300
310-285-7311 (fax)

Blueprint Artist Management[Mngmnt]
5670 Wilshire Boulevard, Suite 2525
Los Angeles, CA 90036
323-330-0333
323-330-0330 (fax)

Bohrman Agency, The
8899 Beverly Blvd., Ste. 811
Los Angeles, CA 90048
(310) 550-5444

Brandt Company, The
15159 Greenleaf St.
Sherman Oaks, CA 91403
(818) 783-7747
BRANDTCO@aol.com

Note: New material can only submitted through a third party referral.

Brillstein-Grey Management
9150 Wilshire Blvd, Ste 350
Beverly Hills, CA 90212
310-275-6135

Broder-Webb-Chervin-Silbermann Agency, The
9242 Beverly Blvd., Suite 200
Beverly Hills, CA 90210
310-281-3400

Buchwald, Don & Assoc.
6500 Wilshire Blvd., Ste. 2200
Los Angeles, CA 90048
(323) 655-7400

Buchwald, Don & Assoc.
10 East 44th St.
New York, NY 10017
(212) 867-1200

Callamaro Literary Agency
427 N. Canon Dr. Ste 202
Beverly Hills, CA 90210
310-274-6783

Careyes Entertainment [Mngmnt]
9000 Sunset Blvd., Suite 800
Los Angeles, CA 90069
310-888-1240
310-888-1243 (Fax)
CareyesEnt@aol.com

Cedar Grove Agency Entertainment
P.O. Box 1692
Issaquah, WA 98027
CedarGroveAgency@juno.com

Celebrity Endeavors [Mngmnt]
230 Park Avenue, Ste 1000
New York, NY 10169
212-730-8683
www.celebrityendeavors.com

Chasin Agency,The
8899 Beverly Blvd., Ste. 716
Los Angeles, CA 90048
310-278-7505
Chic Productions & Management [Mngmnt]
12228 Venice Blvd Suite 499
Los Angeles, CA 90066
310-391-2152

Circle of Confusion (NY) [Mngmnt]
107-23 71st Rd, Suite 300
Forest Hills, NY 11375
718-275-1012

Circle of Confusion (LA) [Mngmnt]
8548 Washington Blvd.
Culver City, CA 90232
310-253-7777

Coast to Coast Talent Group
3350 Barham Blvd.
Los Angeles, CA 90068
(323) 845-9200

Concept Entertainment [Mngmnt]
Attn: David Faigenblum, Eve LaDue
9348 Civic Center Drive, 3rd Floor
Beverly Hills, CA 90210
(310) 276-6177
(310) 276-9477 (fax)
Email: enquiries@conceptentertainment.biz

Coralie Jr. Theatrical Agency
4789 Vineland Ave., Ste. 100
N. Hollywood, CA 91602
(818) 766-9501
Coraliejr@earthlink.net

Creative Artists Agency
9830 Wilshire Blvd.
Beverly Hills, CA 90212
310-288-4545

Creative Convergence [Mngmnt]
11040 Santa Monica Blvd., Suite 200
Los Angeles, California 90025
(310) 954-8480
(310) 954-8481 fax
info@creative-convergence.com

Creative Management Group [Mngmnt]
301 W. 53rd St., Ste 4K
New York, NY 10019
212-245-3250/ 310-888-0082

Delphinius Talent Management [Mngmnt]
Attn: Cynthia Brohas-Gulacsy
10115 Lochcrest Drive
Cincinnati, OH 45231
513-772-3726
delphiniuscbg@aol.com
www.delphinius-management.com

Diverse Talent Group
1875 Century Park East, Ste. 2250
Los Angeles, CA 90067
310-201-6565

Douroux & Co.
815 Manhattan Ave., Ste. D
Manhattan Beach, CA 90266
douroux@relaypoint.net
http://www.relaypoint.net/~douroux

Dytman & Associates
9200 Sunset Blvd, Ste. 809
Los Angeles, CA 90069
(310) 274-8844

Edmonds Management [Mngmnt]
1635 N. Cahuenga Blvd., 5th Floor
Los Angeles, CA 90028
(323) 769-2444

Elkins Ent. Corp [Mngmnt]
8306 Wilshire Blvd, Ste. 438
Beverly Hills, CA 90211
(818) 501-9900
info@elkinsent.com
www.elkinsent.com

Endeavor Agency, The
9701 Wilshire Blvd., 10th Floor
Beverly Hills, CA 90212
(310) 248-2000

Energy (Brooklyn Weaver)[Mngmnt]
999 N. Doheny Dr. #711
Los Angeles, CA 90069
310-274-3440
http://www.energyentertainment.net

Ensemble Entertainment [Mngmnt]
10474 Santa Monica Blvd., Ste. 380
Los Angeles, CA 90025
(310) 882-8900

Envision Management [Mngmnt]
9255 Sunset Suite 300
Los Angeles, CA 90069
310-859-0599

Epstein-Wyckoff-Corsa-Ross & Assoc.
280 S. Beverly Dr., Ste. 400
Beverly Hills, CA 90212
(310) 278-7222 / 212-586-9110

Evatopia Entertainment [Mngmnt]
400 S. Beverly Drive, Ste. 214
Beverly Hills, CA 90212
www.evatopia.com

Exile Entertainment [Mngmnt]
732 El Medio Avenue
Pacific Palisades, CA 90272
(310) 573-1523
Submit queries to: exilesubs@yahoo.com

Fenton Kritzer Entertainment [Mngmnt]
8840 Wilshire Blvd., 3rd Floor
Beverly Hills, CA 90211
(310) 268-2200
(323) 769-2444

Elkins Ent. Corp [Mngmnt]
8306 Wilshire Blvd, Ste. 438
Beverly Hills, CA 90211
(818) 501-9900
info@elkinsent.com
www.elkinsent.com

Endeavor Agency, The
9701 Wilshire Blvd., 10th Floor
Beverly Hills, CA 90212
(310) 248-2000

Energy (Brooklyn Weaver)[Mngmnt]
999 N. Doheny Dr. #711
Los Angeles, CA 90069
310-274-3440
http://www.energyentertainment.net

Ensemble Entertainment [Mngmnt]
10474 Santa Monica Blvd., Ste. 380
Los Angeles, CA 90025
(310) 882-8900

Envision Management [Mngmnt]
9255 Sunset Suite 300
Los Angeles, CA 90069
310-859-0599

Epstein-Wyckoff-Corsa-Ross & Assoc.
280 S. Beverly Dr., Ste. 400
Beverly Hills, CA 90212
(310) 278-7222 / 212-586-9110

Evatopia Entertainment [Mngmnt]
400 S. Beverly Drive, Ste. 214
Beverly Hills, CA 90212
www.evatopia.com

Exile Entertainment [Mngmnt]
732 El Medio Avenue
Pacific Palisades, CA 90272
(310) 573-1523
Submit queries to: exilesubs@yahoo.com

Fenton Kritzer Entertainment [Mngmnt]
8840 Wilshire Blvd., 3rd Floor
Beverly Hills, CA 90211
(310) 268-2200
(310) 385-3139

Festa Entertainment [Mngmnt]
Attn: Dannie Festa
2461 Santa Monica Blvd, Ste. 414
Santa Monica, CA 90404
(310) 315-0569

Firm, The [Mngmnt]
9465 Wilshire Blvd., 6th Floor
Beverly Hills, CA 90212
310-860-8000

FMA Management [Mngmnt]
2270 La Granada Drive
Los Angeles, CA 90068
323-465-5683
E-mail: info@fmamanagement.com
www.fmamanagement.com

Foundation Management [Mngmnt]
1024 N. Orange Dr., Ste. 260
Hollywood, CA 90038
Fax inquiries to: 323-308-3457

Foursight Entertainment [Mngmnt]
8840 Wilshire Blvd., 2nd Floor
Beverly Hills, CA 90211
310-358-3150
310-358-3165 (fax)
E-mail: info@foursight.com
www.foursight.com

Freed Company, Inc., The Barry
2040 Ave. fo the Stars, #400
Los Angeles, CA 90067
310-277-1260
blfreed@aol.com

Freedman Dramatic Agency, Inc., Robert A.
1501 Broadway, #2310
New York, NY 10036
(212) 840-5760

Fusion Management
9200 Sunset Blvd,
Los Angeles, CA 90069
310-278-2888

Gage Group, Inc., The
14724 Ventura Blvd., Suite 505
Sherman Oaks, CA 91403
(818) 905-3800

Gallagher Literary Management [Mngmnt]
Attn: Rob Gallagher
8160 Manitoba, Suite 309
Los Angeles, CA 90293
(310) 822-2070
http://robgallagher.freeservers.com/

Garrick, Dale International
8831 Sunset Blvd, Ste. 941
Los Angeles, CA 90069
(310) 657-2661

Geddes Agency, The
8430 Santa Monica Blvd., Ste 200
Los Angeles, CA 90069
(323) 848-2700

Gelff Agency, The Laya
16133 Ventura Blvd, Suite 700
Encino, CA 91436
(818) 996-3100
Note: They accept QUERY LETTERS from writers but they must include a SASE for reply.

Gersh Agency, Inc., The (L.A.)
232 North Canon Dr, #201
Beverly Hills, CA 90210
(310) 274-6611

Gersh Agency, Inc., The (NY)
41 Madison Avenue, 33rd Floor
New York, NY 10036
(212) 997-1818

GhettoSuburbia Entertainment [Mngmnt]
4335 Van Nuys Blvd., Suite 116
Sherman Oaks, CA 91403
818-749-4920

Gold/Liedtke Agency
3500 W. Olive Ave. #1400
Burbank, CA 91505
818-972-4300

Gordon & Assoc., Michelle
260 S. Beverly Dr, #308
Beverly Hills, CA 90212
(310) 246-9930

Gotham Group, The [Mngmnt]
9255 Sunset Blvd., Suite 515
Los Angeles, CA 90069
(310) 285-0001
(310) 285-0077 (Fax)

Grade A Entertainment [Mngmnt]
368 N. La Cienega Blvd.
Los Angeles, CA 90048
gradeaprod@aol.com

Grant, Savic, Kopaloff and Associates
6399 Wilshire Blvd., Suite #414
Los Angeles, CA 90048
(323) 782-1854
(323) 782-1877 (Fax)
sgkassoc@pacbell.net

Grossman, Larry & Assoc.
211 S. Beverly Dr, #206
Beverly Hills, CA 90212
(310) 550-8127

Gurman, Susan Agency, The
865 West End Ave., #15A
New York, NY 10025
(212) 749-4618

Guy Walks Into A Bar Management [Mngmnt]
7421 Beverly Blvd., Suite #4
Los Angeles, CA 90036
323-930-9935
323-930-9934 (fax)

Handprint Entertainment [Mngmnt]
1100 Glendon Ave., Ste 1000
Los Angeles, CA 90024
310-481-4400

Hartmann Literary, Rogers
6565 Sunset Blvd., Suite 415
Hollywood, CA 90028
323-856-5058

Herman Agency, Richard
124 Lasky Dr, 2nd Floor
Beverly Hills, CA 90212
(310) 550-8913

Himber Entertainment, Inc. [Mngmnt]
211 S. Beverly Dr., Ste. 208
Beverly Hills, CA 90212
(310) 276-2500

Hofflund/Polone [Mngmnt]
9465 Wilshire Blvd., Ste. 820
Beverly Hills, CA 90212
(310) 859-1971

Hollywood View
5255 Veronica St.
Los Angeles, CA 90008
(310) 480-5118
www.hollywood-view.com
Email: dj@hollywood-view.com

More: Inquiry Letters/Emails Only. NO PHONE CALLS.

HSI Entertainment [Mngmnt]
3630 Eastham Dr.
Culver City, CA 90232
310-558-7200

Hung Entertainment Group [Mngmnt]
11040 Santa Monica Blvd., Suite 200
Los Angeles, CA 90025
(310) 954-8480
(310) 954-8481(fax)

IFA Talent Agency
8730 Sunset Blvd., Ste. 490
Los Angeles, CA 90069
(310) 659-5522

Incognito Management [Mngmnt]
345 N. Maple Drive, Ste. 348
Beverly Hills, CA 90210
310-246-1500

Industry Entertainment [Mngmnt]
955 S. Carrillo Dr., Ste. 300
Los Angeles, CA 90048
323-954-9000
FirstnameLastInitial@IndustryEntertainment.com

Infinity Management International [Mngmnt]
7923 Hollywood Blvd.
Los Angeles, CA 90046
323-651-3200
infinitymanagementla@yahoo.com

Inneract Entertainment [Mngmnt]
5410 Wilshire Blvd, Ste. 708
Los Angeles, CA 90036
(323) 965-0967

Innovative Artists
1505 10th Street
Santa Monica, CA 90401
(310) 656-0400 / 212-253-6900

International Creative Mgmt
8942 Wilshire Blvd
Beverly Hills, CA 90211
(310) 550-4000

International Creative Mgmt
40 W. 57th St.
New York, NY 10019
(212) 556-5600

International Digital Artists [Mngmnt]
4223 Glencoe Ave., Ste. C-107
Marina Del Rey, CA 90292
(310) 306-4001

Kallen, Leslie B. Agency
115760 Ventura Blvd, #700
Encino, CA 91436
(818) 906-2785
http://www.lesliekallen.com

Kaplan/Perrone Entertainment [Mngmnt]
(Aaron Kaplan & Sean Perrone)
10202 W. Washington Blvd.
Astaire Bldg. Suite #3003
Culver City, CA 90232
310-244-6681
310-244-2151 (fax)

Kaplan-Stahler-Gumer Agency, Inc.
8383 Wilshire Blvd, Ste. 923
Beverly Hills, CA 90211
(323) 653-4483
ksagency@aol.com

Kersey Management [Mngmnt]
Attn: Andrew Kersey
1409 North Alta Vista Blvd, #304
Los Angeles, CA 90046
323-850-8818

Ketay Agency, The Joyce
1501 Broadway, Ste. 1908
New York, NY 10036
212-354-6825

Kjar Agency, Tyler
5144 Vineland Ave.
N. Hollywood, CA 91601
(818) 760-0321

Klane Company, The Jon [Mngmnt]
120 El Camino Dr, #112
Beverly Hills, CA 90212
(310) 278-0178
http://www.klaneco.com

Lake Agency, Candace The
9200 Sunset Blvd, #820
Los Angeles, CA 90069
(310) 247-2115
clagency@earthlink.net

LaSalleHolland [Mngmnt]
1650 Broadway, Suite 508
New York, NY 10019
(212) 541-4443
(212) 581-6906
www.lasalleholland.com

LMR [Mngmnt]
(Lasher, McManus & Robinson)
1964 Westwood Blvd. #400
Los Angeles, CA 90025
(310) 446-1466
lmrla@earthlink.net

Lantz Office, The
200 W. 57th St. Ste 503
New York, NY 10019
212-586-0200

Lenny, Jack Assoc. (L.A.)
9454 Wilshire Blvd, #600
Beverly Hills, CA 90212
(310) 271-2174

Lighthouse Entertainment [Mngmnt]
409 N. Camden Dr. #202
Beverly Hills, CA 90210
(310) 246-0499
Lovett Management [Mngmnt]

1327 Brinkley Ave.
Los Angeles, CA 90049
(310) 451-2536

** Please note that we do our best to keep the page updated and accurate, but addresses can frequently change (without our knowing it) so it is best to check with the company first to make sure they have not moved.

Magnet Management
6380 Wilshire Blvd. Suite 915
Los Angeles, CA 90048
(323) 658-8095
(323) 658-8636 (fax)

Major Clients Agency
8530 Wilshire Ave., 3rd Fl.
Beverly Hills, CA 90211
(310) 967-0200

Management 360 [Mngmnt]
9111 Wilshire Blvd.
Beverly Hills, CA 90210
(310) 272-7000

MBST Entertainment [Mngmnt]
345 N. Maple Drive, Ste. 200
Beverly Hills, CA 90210
(310) 385-1820

Media Artists Group
6404 Wilshire Blvd., Ste 950
Los Angeles, CA 90048
(323) 658-5050

Messina Baker Entertainment Corp. [Mngmnt]
955 Carrilo Dr., Ste. 100
Los Angeles, CA 90048
(323) 954-8600

Metropolitan Talent Agency
4526 Wilshire Blvd.
Los Angeles, CA 90010
(323) 857-4500

Mocean Pictures [Mngmnt]
1680 N. Vine St., Suite 811
Hollywood, CA& 90028
323-466-5056
Query@MoceanPics.com
www.MoceanPics.com

Modus Entertainment [Mngmnt]
110 South Fairfax Avenue, Suite 250
Los Angeles, CA& 90036
323-930-6730
323-930-6736 Fax

Monroe Pritchard Monroe Literary Agency
722 Ridgecreek Drive
Clarkston, GA 30021
404-296-4000
allinline@hotmail.com

Muraviov Company, The [Mngmnt]
Attn: Kathy Muraviov
10607 Montgomery Avenue
Granada Hills, CA 91344
818-830-3449/818-425-5165
TheMuraviovCo@aol.com

Narelle Sheehan Management-NSM [Mngmnt]
PO Box 5055
Beverly Hills, CA 90209
(310) 385-0419

Niad Management [Mngmnt]
3465 Coy Drive, Suite 210
Sherman Oaks, CA 91423
(818) 981-2505
(818) 386-2082 Fax
www.niadmanagement.com

Nick Yellen Creative Agency
1501 Broadway Ste. 2310
New York, NY 10036
(212) 840-5760
(212) 840-5776 Fax

NYCAgency@nyc.rr.com
Nine Yards Entertainment [Mngmnt]
8530 Wilshire Blvd., 5th Floor
Beverly Hills, CA 90211
(310) 289-1088
(310) 289-1288 Fax

O/Z Management [Mngmnt]
c/o Phoenix Pictures
10202 W. Washington Blvd., Ste. 115B
Culver City, CA 90232
(310) 244-6380

Orange Grove Group, Inc., The
12178 Ventura Blvd, #205
Studio City, CA 91604
(818) 762-7498
gregmayo@orangegrovegroup.com

Original Artists
9465 Wilshire Blvd., Ste 840
Beverly Hills, CA 90212
(310) 275-6765

Oscard, Fifi Agency, Inc.
110 W. 40th St., 16th Floor
New York, NY 10018
(212) 764-1100
212-840-5019 fax
FifiOscard@aol.com
www.fifioscard.com

Overland Literary Management [Mngmnt]
1701 Harvard Street, NW
Washington, DC 20009
202/667-1135
310/205-8998 (Los Angeles)
OverlandLitMgt@prodigy.net

aradigm
360 N. Crescent Drive, North Bldg.
Beverly Hills, CA 90210
(310) 288-8000
(310) 288-2000 fax

Perelman, Barry Agency, The
1155 No. La Cienega Blvd. Suite 412
West Hollywood, CA 90069
310-659-1122

PINCH Entertainment [Mngmnt]
Attn: Shelley Spevakow
7361 Rosewood Avenue
Los Angeles, CA 90036
323-782-3815
info@pinchent.com
www.pinchentertainment.com

Pitt Group, The [Mngmnt]
9465 Wilshire Blvd,. Ste. 480
Beverly Hills, CA 90212
310-246-4800

Pleshette Literary Agency, The Lynn
2700 N. Beachwood Dr.
Hollywood, CA 90068
(323) 465-0428

Preferred Artists
16633 Ventura Blvd., #1421
Encino, CA 91436
(818) 990-0305

Preminger, Jim Agency, The
450 N. Roxbury Dr. PH 1050
Beverly Hills, CA 90210
310-860-1116

Principato-Young Management [Mngmnt]
The Brill Bldg, 1619 Broadway, 9th Fl
New York, NY 10019
212-603-1843 / 310-274-4474

Privilege Talent Agency
14542 Ventura Blvd., Suite 209
Sherman Oaks, CA 91403
(818) 386-2377
(818)386-9477 fax

Pro and Con Productions Talent Management [Mngmnt]
PO Box 18376
Encino, CA 91416-8376
(818) 973-2282
proandconprods@geocities.com
http://www.geocities.com/proandconprods

Professional Artists
321 W. 44th St., Ste. 605
New York, NY 10036
(212) 247-8770

Propaganda Management [Mngmnt]
c/o Propoganda Films
1741 Ivar Ave.
Hollywood, CA 90028
323-802-6800

Pucket Literary
945 N. Pasadena, #80
Mesa, Arizona 85201
(480) 964-4807

Radmin Company, The [Mngmnt
9201 Wilshire Blvd., Ste. 305
Beverly Hills, CA 90210
(310) 274-9515

Renaissance Literary/Talent Agency
A division of Artist's Management Group
9465 Wilshire Boulevard
Beverly Hills,CA 90212
(310) 860-8000
renaissance@earthlink.net

Renegade Management
A Crystal Sky Company
1901 Avenue of the Stars, Suite 605
Los Angeles, CA 90067
(310) 843-0223 x 37
(310) 553-9895
Mail@RenegadeMgt.com

Rigberg-Roberts-Rugolo Company, The [Mngmnt]
1180 S. Beverly Dr., Ste. 608
Los Angeles, CA 90035
(310) 712-0712

Roberts Company, The [Mngmnt]
3000 W. Olympic Blvd, Bldg. 2 Ste. 2525
Santa Monica, CA 90404
310-264-4238

Rogers & Associates, Stephanie ([Mngmnt])
8737 Carliatas Joy Court
Las Vegas, NV 89117
702-255-9999

Rogue Pictures ([Mngmnt])
10202 W. Washington Blvd Gene Autry Bldg. Rm 1019
Culver City, CA 90232
310-244-7399

Rose Agency, Brant
6671 Sunset Blvd., Suite 1584B
Los Angeles, CA 90028
323-460-6464

Rosenberg Office, The Marion [Mngmnt]
PO Box 69826
Los Angeles, CA 90069
(323) 822-2793

Rothman Agency, The
9465 Wilshire Blvd, Ste. 840
Beverly Hills, CA 90212
(310) 247-9898

Rozon/Mercer Management [Mngmnt]
201 N. Robertson Blvd., Ste. F
Beverly Hills, CA 90211
310-777-1100

Sarkes/Kernis Company, The [Mngmnt]
315 S. Beverly Drive, Ste. 216
Beverly Hills, CA 90212
310-785-0444

Schechter Company, The Irv
9460 Wilshire Blvd, Ste. 300
Beverly Hills, CA 90212
(310) 278-8070

Schwartzman Office, The Paul
11777 San Vicente Blvd.
Los Angeles, CA 90049
(310) 828-4040

Seifert Management, Linda [Mngmnt]
22 Poland Street
London, England W1F 8QQ
+44-20-7292-7390
+44-20-7292-7391 (fax)
E-mail: contact@lindaseifert.com

Seven Summits Pictures and Managment [Mngmnt]
8906 West Olympic Boulevard, Garden Suite
Beverly Hills, CA 90211
310-550-6777
310-550-0606 (fax)

Shapira & Assoc., David
15821 Ventura Blvd, Ste 235
Encino, CA 91436
(818) 906-0322

Shapiro-Lichtman, Inc.
8827 Beverly Blvd.
Los Angeles, CA 90048
(310) 859-8877

Shapiro/West & Associates, Inc. [Mngmnt]
141 El Camino Dr., Ste. 205
Beverly Hills, CA 90212
310-278-8896

Shuman Company, The [Mngmnt]
3815 Hughes Ave., 4th Fl
Culver City, CA 90232
310-841-4344
shumanco@aol.com

Siegel Management, Rick [Mngmnt]
8060 Melrose Ave., 4th Floor
Los Angeles, CA 90046
(323) 852-1776

Simon Entertainment, Andrea [Mngmnt]
14011 Ventura Blvd., Suite 101-S
Sherman Oaks, CA 91423
818-380-1901
818-380-1932 (fax)
info@andreasimonent.com

Sladek Entertainment Corp., Daniel (Daniel Sladek)
8306 Wilshire Blvd. #510
Beverly Hills, CA 90211
323-934-9268
323-934-7362 (Fax)

www.danielsladek.com
SMA, LLC
PMB #615
9899 Santa Monica Boulevard
Beverly Hills, CA 90212
310-203-8787
e-mail: agentsara@flashcom.net
website: http://www.smaagency.com

Stein Agency, The
5125 Oakdale Avenue
Woodland Hills, CA 91364
818-594-8990
818-594-8998 fax
mail@thesteinagency.com

Steinberg Talent Management Group [Mngmnt]
1600 Broadway, Ste. 410
New York, NY 10019
(212) 582-7589
nycomedy@aol.com
http://members.aol.com/nycomedy

Stone Manners Agency, The
6500 Wilshire Blvd #550
Los Angeles, CA 90048
(323) 655-1313

Summit Talent & Literary Agency
9454 Wilshire Blvd., Ste. 203
Beverly Hills, CA 90212
310-205-9730

TalentScout Management
12228 Venice Blvd., Suite 539
Los Angeles, CA 90066
310-281-6213
310-397-3695 (Fax)
contact@atalentscout.com
http://www.atalentscout.com/

Tandem Entertainment Management [Mngmnt]
7095 Hollywood Blvd., #1358
Hollywood, CA 90038
(323) 860-7816
tandemmgmt2002@yahoo.com

Targ Literary Agency, Roslyn
105 W. 13th St., Ste. 15 - E
New York, NY 10011
(212) 206-9390
roslyntarg@aol.com

Tavel Entertainment [Mngmnt]
9171 Wilshire Blvd., Ste. 406
Beverly Hills, CA 90210
310-278-6700

Thruline Entertainment [Mngmnt]
9250 Wilshire Blvd., Ground Floor
Beverly Hills, CA 90212
310-595-1500

Tollin/Robbins Management
10960 Ventura Blvd, 2nd Fl
Studio City, CA 91604
818-766-5004

Tudor Management Group [Mngmnt]
4712 AdmiraltyWay, #560
Marina del Rey, CA 90292
(310) 247-1660
info@tudorgroup.com

Turtle Agency, The
7720B El Camino Real, #125
Carlsbad, CA 92009
760-632-5857
cturtlewal@aol.com

United Talent Agency
9560 Wilshire Blvd, 5th floor
Beverly Hills, CA 90212
(310) 273-6700

Unlimited Management
1627 Pontius Ave, Ste 200
Los Angeles, CA 90025
310-914-9171

Vision Art Management
9200 Sunset Blvd. PH 1
Los Angeles, CA 90069
310-888-3288

Visionary Management [Mngmnt]
8265 Sunset Blvd., Ste. 203
Los Angeles, CA 90046
(323) 848-9538
tparz@aol.com

Warden, White & Associates
8444 Wilshire Blvd., 4th Floor
Beverly Hills, CA 90211
(323) 852-1028

Wauhob Agency, Donna
3135 Industrial Road, Ste. 204
LLas Vegas, NV 89109
(702) 733-1017
dwauhob@aol.com

Waugh Talent Agency, Ann
4741 Laurel Canyon Blvd, Ste. 200
N. Hollywood, CA 91607
(818) 980-0141

Wax Agency, The
120 El Camino Dr., Ste. 114
Beverly Hills, CA 90212
(310) 550-8738

William Morris Agency
One William Morris Place
Beverly Hills, CA 90212
310-859-4000 / 212-586-5100

Zebra Agency
Broadlands House, 1Broadlands,Shevington
Lancashire, England WN6 8DH
0794 958 4758
www.zebraagency.co.uk
Email: mail@zebraagency.co.uk

Notes: Non-fiction and general fiction including crime, suspense and drama, murder, mysteries, adventure, thrillers, horror, and science fiction. Plus scripts for TV/radio/film/theatre. No unsolicited material; Send preliminary letter giving publishing history and brief cv (resume), with ten pages of script (plus S.A.E.). No phone calls or submissions by fax or e-mail.

MULTICULTURAL SITES

http://www.salesvantage.com
http://www.extrasformovies.com/audition_acrhives1.asp
www.diversityworking.com/career/Arts_and_Entertainment/Management,_business,_and_financial_agents/Talent_Agent.html
http://www.ethnicvogue.com/
http://directory.jayde.com/entertainment_industry_employment.html

PROFESSIONAL TRAINING FOR KIDS/TEENS

www.actingconnection.com

Hollywood Actors Studio

Pursuing a career in the arts takes a lot more than strong enthusiasm or even a determined attitude. It takes courage, vision & tenacity; but above all else it takes faith & confidence in yourself.

Good performers understand the importance of developing their talent to the fullest and constantly honing their skills. Strong coaching is at the heart of the matter of many successful careers. Private one-on-one intensive training provides you with the unparalleled experience of having a master coach custom-tailor your development for your skills and abilities. Whether in Los Angeles or in your hometown, private training is an invaluable resource for actors and performers at every level of their development.

Hollywood Actors Studio is a safe and creative professional medium for all aspiring actors, performers, public speakers & voice-over artists dedicated to developing solid skills and gaining confidence as artists and individuals.

We offer a variety of solid skill development programs at all levels to fine tune your instrument and grow your talent. These programs cover a wide & complete spectrum of distinctions and techniques essential in the pursuit of self-expression as an art form and a career.

Human beings need to communicate their inner feelings as a function of their aliveness & well-being. To do it publicly & artistically while contributing to the lives of others is a precious experience. The studio's aim is to empower individuals to reach inside and find the necessity to share & express themselves fully. Hollywood Actors Studio successfully contributes in developing strong, independent and free-spirited artists.

www.proactorsstudio.com
Classes taught by Nick Conti.

The goal of the **Professional Actor's Studio** is to prepare actors for all levels for stage, film, television, and commercial auditions so that you get the call back *and* the part. To that end, the acting techniques used by Nick Conti's Professional Actor's Studio are a well balanced mixture of old and new -- from Stanislavski to Meisner, from method acting to "new school" acting exercises.

Our team of instructors brings a diverse array of experiences -- mixed with honest professionalism -- to give you the best acting school that can be found in Metro Atlanta.

Using the Nick Conti script analysis techniques and your own creativity, we prepare you to give an audition that is not just *good* -- we prepare you to give an audition that is truly *special* and *magical*. Using our techniques, you stand out in the crowd. Our acting exercises help you to become a working actor!

Whether you are a new actor just beginning to understand the "business of show business" or a seasoned, experienced actor trying to stay sharp -- we give you clear, specific, useable feedback.

All classes are ongoing, so you're constantly working and honing your skills. The Professional Actor's Studio is always here. You can take a break from the acting school to work on a project and then, when the gig is complete, return to acting classes to continue your education.

Find out for yourself why the Professional Actor's Studio is the best, most recommended acting school classes in Atlanta. We invite you to call us to arrange an appointment to audit a free acting class.

www.youract.tv

An exciting feature of studying at **yourACT** is the chance to benefit from an industry pro's observations, opinions, and advice. ***Ongoing courses include visits with one of Atlanta's top agents or casting directors, answering questions and giving you specific feedback on your acting skills.***

We offer a variety of on-camera Atlanta acting classes for all levels of Kids, Teens, and Adults:

Beginners learning the basics

Actors continuing their training and improving their audition skills

Advanced actors looking to stay sharp

You'll receive in-depth training, honest critique, and constructive suggestions from your instructor as you enhance your abilities and build your confidence. Your training is only the beginning. At yourACT you have the opportunity to meet and show your work to an agent or casting director in an intimate, informal setting. You'll gain valuable tools for success whether you're currently seeking representation or if you already have an agent and are actively auditioning. Receive inside tips about the business, get an agent's or casting director's take on your audition skills, and hear their personal advice on your work and what to do next. Advanced actors perform live for agents, casting directors, and industry pros that direct and critique them.

www.actingzone.com

ACTING COACHES, TEACHERS, SCHOOLS DIRECTORY NATIONWIDE

www.redbirdstudio.com

This is the place for beginning actors and actresses to learn about acting and the acting business. For actors and actresses aspiring to become professional or for those who want to take their avocation to the highest artistic level possible. **Acting workshops on-line**

TALENT UNIONS

www.actorsequity.org

Actors' Equity Association

Actors' Equity Association is the labor union representing over 45,000 American actors and stage managers working in the professional theatre.

For 90 years, Equity has negotiated minimum wages and working conditions, administered contracts, and enforced the provisions of our various agreements with theatrical employers across the country.

Equity publishes a booklet that describes its history, functions, and services.

Booklet available online

www.aftra.org

American Federation of Television and Radio Artists:

The American Federation of Television and Radio Artists (AFTRA) is a national labor union representing over 70,000 performers, journalists and other artists working in the entertainment and news media.

AFTRA's scope of representation covers broadcast, public and cable television (news, sports and weather; drama and comedy, soaps, talk and variety shows, documentaries, children's programming, reality and game shows); radio (news, commercials, hosted programs); sound recordings (CDs, singles, Broadway cast albums, audio books); "non-broadcast" and industrial material as well as Internet and digital programming.

AFTRA's membership includes an array of talent - sound recording membership includes artists who bring pop, rock, country, classical, folk, jazz, comedy, Latin, hip hop, rap and R&B to the world. AFTRA members perform in television and radio advertising, non-broadcast video, audio books and messaging, and provide their skills for developing technologies such as interactive games and Internet material.

Contract Negotiation and Enforcement

The union negotiates and enforces over 300 collective bargaining agreements that guarantee minimum (but never maximum) salaries, safe working conditions and health and retirement benefits. When the union is unable to resolve disputes with employers, AFTRA contracts include procedures for binding arbitration. The cost of these proceedings is paid by the union. In the past year, AFTRA pursued claims, grievances and arbitration that recovered over $10 million for AFTRA members.

Advocacy

AFTRA advocates on legislative and public policy issues that directly affect members' wages and working conditions. This includes lobbying and support for such issues as ownership consolidation in the broadcast industry, Equal Employment Opportunity laws and regulations, Copyright and performance rights issues, "non-compete" restrictions, OSHA and local safety regulations, and more. AFTRA also represents artists and journalists in judicial proceedings, often as *amicus*, and before the U.S. Copyright Office.

Member Benefits

AFTRA supports the welfare of its members through benefits and programs sponsored or supported by the union. AFTRA was the first industry union to establish employer-paid health and retirement plans for members and their dependents that qualify. The Health Fund provides comprehensive medical and hospital benefits, a dental plan, prescription drug program, life insurance and mental health and substance abuse programs. The Retirement Fund is portable, so that wherever members work under an AFTRA contract -- whether on a television or radio program, a sound recording, commercial or industrial show -- employer contributions are made on their behalf. For more detailed information about the AFTRA Health & Retirement Funds, visit its web site at www.aftrahr.com.

AFTRA membership also provides access to a series of supplemental benefits, discounts and other programs through groups such as Group Benefits Associates, TEIGIT, UnionPlus/Union Privilege and MusicPro.

The AFTRA Foundation, a separate tax-exempt organization funded entirely by voluntary contributions, works in even broader ways to benefit performers and broadcasters and supports the educational and charitable objectives of AFTRA.

Scholarships are available to AFTRA members and dependents through the AFTRA Heller Memorial Foundation. AFTRA members are also eligible for assistance through organizations such as Theater Authority and the Actors Fund. In addition, several AFTRA Locals sponsor scholarship programs and various charitable foundations for the benefit of AFTRA members and/or their dependents.

Governance and Administration

AFTRA was founded and governed by members. Each of the Locals around the country elects its own Board of Directors, in addition to electing representatives to the National Board of Directors. Those Local and National Boards set the direction of the organization and establish its policies on the local and national levels respectively. AFTRA also holds a National Convention every two years, and delegates are directly elected on a proportional basis from each Local to represent the membership at the Convention. All Board members and Officers -- both Local and National -- serve without pay. Policies and programs developed by the elected members are implemented and administered by salaried staff at the national and local level.

AFTRA's current National President is Los Angeles actor John Connolly. Its senior administrator and labor relations professional is National Executive Director Greg Hessinger.

Other Initiatives

AFTRA actively supports equal employment opportunities for minorities and women in the broadcast and entertainment industries, and its contracts call for diversity and fairness in hiring practices. AFTRA also protects its members by franchising talent agents who represent performers in its jurisdictions.

AFTRA Local offices offer a wide variety of special programs tailored to the needs of members who live and work in their area. These programs range from publication of talent guides distributed to local producers, agents and casting directors; scheduling of conservatories or seminars which help members improve skills, write resumes and develop audition techniques; opportunities for members to meet casting directors, agents and others who hire talent; and local showcases to highlight new talent to the industry.

Affiliations

- AFL-CIO
- Department for Professional Employees of the AFL-CIO (DPE)
- Associated Actors and Artistes of America (4-A's)
- International Federation of Actors (FIA)
- International Federation of Journalists (IFJ)

www.sag.org

Screen Actors Guild

Welcome to Screen Actors Guild, a labor organization whose mission is to negotiate the best possible wages and working conditions for professional actors – even young ones.

The following pages present an overview of some of the information you will need to make your way through this exciting, but often challenging, industry.

If guided carefully, a young performer's acting career can give them a head start on their adult lives, help them develop leadership skills, teamwork skills, self-confidence and the ability to earn income.

We call this industry "show – business" for a good reason – it is a business. You should treat it as such. That being said, it should also be fun. If for any reason, it is not fun and educational for you, get out. If your goal is to make a lot of money and get really famous, really quickly, your hopes will probably never be recognized. You should be in this business only if you are truly dedicated, having plenty of stamina, and a passion for the craft of acting.

Screen Actors Guild does its utmost to protect young performer's safety and wages; however it is always your parent or guardian's responsibility to insure that you are being treated fairly and with respect.

HOMESCHOOLING FOR ENTERTAINERS

www.keystonehighschool.com

KEYSTONE National High School   Contact us at 800-255-4937

Keystone: an experienced leader, a dependable partner.

For over thirty years, Keystone National High School has been recognized as an innovator in home education and distance learning approaches. Families across the country and around the world have partnered with Keystone and its staff to help prepare their high school students for a successful future.

- Our Advantage
- Programs
- Curriculum
- Tuition
- Community

Keystone National High School is a division of KC Distance Learning, Inc. (KCDL), the leading provider of distance learning programs for the secondary school market. KCDL uses the power of distance learning to bring quality educational opportunities within reach of every student. In addition to Keystone National High School, KC Distance Learning operates the following programs.

Keystone Credit Completion Program The Credit Completion Program (originally founded as The Learning and Evaluation Center) has long been the number one choice for school officials seeking a proven, successful credit completion program for their students. Since 1974, over 160,000 students from 5,000 schools have used its independent study subject extensions to successfully complete courses they had previously failed.

www.zionacademy.com

5220 176th SW 12   Lynnwood, WA 98037   Ph/Fax: 425-743-4300   zionacademy@verizon.net

Zion Academy of America is an established private school with a complete distance learning homeschool educational program serving America's families nationwide since 1985. It is dedicated to the education of children for grades K-12. Enrolling in Zion Academy of America students can benefit from a home education and take advantage of being enrolled in a **fully accredited** private school. It uses the highest quality educational curriculum and materials.

Our dedication is to parents by helping them achieve the best home education possible for their children. Our goal is to help one student at a time to succeed in school and feel better about themselves at the same time.

There are many programs for homeschooling, many curriculums, and many places that you can acquire homeschool materials. But Zion Academy is **UNIQUE** because it is an **established private school**, and all of its students are homeschooled. This is an unusual concept, but it works!!!

This is the reason that thousands of students have enrolled in Zion Academy over the years. Protection of the student has been the main reason that we have gone through a major undertaking for Zion Academy to be properly established and nationally recognized as a private school. You will find it a privilege to be enrolled in Zion

Academy and that your child's credits will be recognized and accepted by the next school. All graduating Seniors receive a High School Diploma.

Zion Academy was established in 1985 and has successfully assisted thousands of families and students across America with its tried and proven program. Zion Academy is privately owned and operated by experienced professionals in the field of education.

Zion Academy of America is a division of Mount Zion Ministries International Inc., an interfaith ministry. Zion Academy is registered with the American College Testing Program (A.C.T.) and the College Entrance Examination Boards (SAT). Zion Academy is fully accredited by the National Association of Private Schools.

www.his.edu

HOME STUDY INTERNATIONAL

P.O. Box 4437 Silver Spring, MD 20914-4437
Phone 301-680-6570   Fax 301-680-5157

History
At the turn of the century, educational opportunities were rare. Correspondence education was increasing in popularity within the United States. An educator by the name of Frederick Griggs envisioned educating people around the world. His vision took shape in the establishment of The Fireside Correspondence School in 1909. The goal was to provide the benefits of an education to those unable to attend traditional schools.

Within two years, The Fireside Correspondence School (later renamed Home Study Institute and more recently Home Study International) was offering 11 secondary and 9 college courses. By 1916 its students represented nearly every state and province in North America, as well as 10 other countries. Since then, more than 235,000 people have studied with HSI.

HSI is part of the Seventh-day Adventist school system. Today HSI plays a unique and vital role in the educational development of students of all ages in all parts of the world. In 1990 the HSI Board of Directors assigned names to its three academic divisions. Thus Home Study Elementary School, Home Study High School and Griggs University (GU) became a part of HSI's terminology. In 1991, GU began offering college degrees to international students who have no access to an Adventist campus.

People from all walks of life have discovered that the quiet conditions of private correspondence study help develop self reliance, independent thinking and responsibility. From its early years when it operated out of a one-room office, HSI has developed into a worldwide school that maintains high scholastic standards and utilizes the services of qualified professionals in all phases of its operation. HSI emphasizes the personal touch in its student-teacher relationships.

Accreditation: Middle States Association for Colleges and Schools HSI's K-8 program is regionally accredited by MSAl, Southern Association of Colleges and Schools HSI's K-12 program is regionally accredited by SACS, Accredited by the Accrediting Commission of the Distance Education and Training Council, a national accrediting agency recognized by the U.S. federal government, Accredited by the Accrediting Association of Seventh-day Adventist Schools, Colleges, and Universities, Approved by the Maryland State Department of Education for Kindergarten, Elementary, and High School, Member of the American Council on Education, Member of The American Association of Collegiate Registrars and Admissions Officers, Member of International University Consortium, Member of the International Council for Distance Education, Member of University Continuing Education Association, and Member of Adventist Colleges Abroad.

www.ihomeschool.com

IHOMESCHOOL

ihomeschool has four solutions to engage in web-based homeschooling. Each solution offers curriculum for your child to complete one grade level in one year. Our programs allow you to customize your child's education to fit your family's needs.

1. The **Core Curriculum**, an entry level package, includes curriculum, workbooks, email and access to message boards used to interact with other homeschooling families.

2. The **Achievers** package includes everything from the *Core Curriculum*, as well as comprehensive study aids such as the World Book Online web-based encyclopedia, online professional tutors, a video library with more than 26,000 educational video clips, and dozens of educational links.

3. The **Advanced Achievers** adds an interactive teacher-led class once a week for two courses. Also, students have access to teaching assistance throughout the week.

4. The **Performance Plus** package is your complete homeschooling solution. Added to the features available with the *Advanced Achiever* program is the option to learn a foreign language. We offer courses in French and Spanish. With **Performance Plus**, we will provide teaching services that may include online classes once a week for each course. Teachers may also review or assess student work and support the parent educator with teaching tips and guidance.

Online organizational tools and educational resources assist your homeschooling experience in our innovative solution. Supplement any curriculum with the **Reference Desk & Learning Management System (LMS)** solution.

For those homeschooling families searching for one course to complete their curriculum, our **Single Course** offerings are just the solution!

You don't have to be a computer expert; ihomeschool is a turnkey solution—just follow our easy step-by-step process. For more information, give us a call at **866-374-6640**.

Easy Monthly Payments

ihomeschool offers convenient payment options via our secure online ordering system. You may make your payment in full, or you may use our simple monthly payment plan based on the ihomeschool solution that you choose.

System Requirements

Our electronic curriculum and Video Library are driven by the latest technology; therefore we do suggest the following minimum system requirements:

- Windows XP
- Pentium IV
- 512 MB RAM, 1GB preferred
- 1 GB Hard Drive
- Internet Explorer 5.5 or newer
- Firefox 1.3 or newer
- Safari 1.2 or newer
- Windows Media Player or QuickTime

- Macromedia Shockwave
- Adobe Acrobat Reader

- Complete Packages
    - Pre-Kindergarten
    - Kindergarten
    - Grades 1–3
    - Grades 4–6
    - Grades 7–8
    - High School
- Single Courses
    - Grades 1–3
    - Grades 4–6
    - Grades 7–8
    - High School
- Reference Desk
  & LMS

www.calvertschool.org

(888) 487-4652

CALVERT SCHOOL

Best education

Calvert Educators selected the best educational resources and have woven them together to create an experience that is more than the sum of its parts. Our **integration of material** from subject to subject reinforces your instruction and makes for a more engaging educational experience. Our **time-tested**, academically stimulating, and accredited program includes online activities. We also offer virtual classes and correspondence testing services for additional support. Our complete program provides you with the peace of mind of an education with **no gaps** in instruction.

Best support

Your personal attention is complemented by our **personal attention**. Only Calvert provides Education Counselors to answer your academic and homeschooling questions. Our customer service staff is an easy phone call away. Payment plans and re-use programs can help ease financial concerns.

Best fit

Placement **flexibility** in math, Calvert Academic, or Calvert Scholastic programs positions your child for a successful year. We offer enrichment courses that are designed to work individually or to reinforce our core instruction.

Best use of time

Our **complete** program includes access to Education Counselors, **detailed plans** in the Lesson Manual, answer keys for daily work, and all of the materials you will need--textbooks, workbooks, and supplies--to make the most of your instructional time.

www.corecurriculum.com

CORE CURRICULUM OF AMERICA

14503 S. Tamiami Trail
North Port, Florida 34287
1-888-689-4626
Homeschool Curriculum
Packages Customized
For YOU!

After receiving input from both the student and parent we design a special, inexpensive, innovative curriculum package for your child that assures *success*! A homeschool curriculum package *Customized* for your child based on his/her abilities, needs, and desires - to help motivate, challenge, and most importantly......*enrich* your child's home learning experience.

INTERNATIONAL CHILD INSURANCE

www.childinsurance.com

CHILD INSURANCE

The best time to start life insurance coverage for your children is when they are young and in good health. Globe Life can provide guaranteed life insurance protection for your children or grandchildren. The Young American Plan is children's insurance that is a whole life policy. The policy can be the beginning of a financial estate and is a smart way to protect your child's ability to get life insurance coverage in the future.

When considering children's insurance, it's important to find an insurance company that has adequate financial strength. Globe Life continues to receive one of the highest national industry ratings for insurance carriers - **A+ Superior****. This rating is the second-highest awarded out of thirteen possible ratings based on the latest analysis of Globe Life's financial strength, management skill and integrity.

* Rates after the first month can be found during the full rate quote process

** A.M. Best Company - independent insurance analysts since 1899. This rating refers only to the overall financial status of the company and is not a recommendation of the specific policy provisions, rates or practices of the insurance company.

www.norfolkgrp.com

NORFOLK INTERNATIONAL GROUP

"We offer our clients a portfolio of diverse international insurance products for North Americans working in other countries, as well as foreign workers temporarily employed in Canada or the United States. Our clients also have access to our many years of insurance industry experience, our affiliations with wholesale risk-taking markets in several countries, and our substantial expertise in the development and administration of customized international benefits programs."

"Since 1980, The Norfolk Group has been building partnerships with companies around the world by providing exceptional service and expertise in benefits consulting, program design and implementation. Whether you are looking for a program design or outsourcing service, we bring a well-informed global perspective and creative problem solving techniques to your doorstep. Our goal is to work with you to develop flexible, cost-effective programs and services that make sense for your company and your employees."

Ri Norfolk International is North America's leading provider of worldwide customized expatriate insurance solutions. We are proud of our successes in developing customized cost-effective plans to meet virtually any situation.

Our international services are diverse and range from basic travel insurance to all of the benefits and services required for long term expatriate assignments. We can meet the needs of North Americans working in other countries, as well as foreign workers temporarily employed in Canada or the United States.

Our international insurance includes:

- Medical/Dental Plans

- Life/Accidental Death & Dismemberment
- Short-Term and Long-Term Disability
- Health Care Spending Accounts
- Employee Assistance Programs
- Commercial General Liability

Kidnap and Ransom **chard Albert, BBA – President**

Unique solutions
Norfolk International is widely recognized as a leader in the delivery of expatriate benefit solutions to all types of organizations. Since 1980, Norfolk has been creating and customizing benefit options that address the particular needs of our global clients. As a result, the products we offer also include:

Critical Illness
Business Travel
International Payroll
Bridging LTD

Insurers we use
The insurers we use are the finest in the world. They include:

Lloyd's - The world's leading insurance market, including Odyssey™ insurance for individuals. Hauteville - Providing expatriate insurance for more than 40 years, including Envoy insurance for individuals. Allianz - Global risk insurance. AIG - The world's leading U.S.-based international insurance and financial services organization. Cigna International - Expatriate benefits for multinational employer groups and international business travelers. Aetna – Comprehensive global health coverage. Zurich International – one of the world's leading providers of international insurance and investment products.

Calgary Head Office
Suite 1100
940 - 6 Avenue SW
Calgary, Alberta
Canada T2P 3T1
Tel: (403) 232-8545
Fax: (403) 265-9425
Toll free: 1-800-672-6089

Toronto Office
Unit 13
1845 Sandstone Manor
Pickering, Ontario
Canada L1W 3X9
Tel: (905) 420-5707
Fax: (905) 420-7042

Houston Office
Norfolk Global Benefits Management LP
5210 Evergreen Valley Drive
Kingwood, Texas
USA 77345
Tel: (281) 782-5636

norfolk@norfolkgrp.com

INTERNATIONAL BANKING

www.officeshorecorporation.com

Offshore Banking - Asset Protection and Privacy

The objective of **offshore banking** is asset protection and financial privacy. At Offshorecorporation.com we specialize in offshore bank accounts, including Swiss bank accounts, as well as offshore corporations, and offshore merchant accounts. We can open both personal **offshore** and corporate accounts. By going offshore you can benefit from financial secrecy and lawsuit protection.

To obtain maximum asset protection and privacy, the best advice is to establish an offshore corporation to own your offshore account. Having an offshore account in your own name raises little question that the account is yours. All account deposits and wire transfers to the offshore bank would be in your name, leaving a trail. With offshore, having the offshore account titled to your offshore corporation, your relationship with the offshore bank remains private. Any deposits or wire transfers are made to the offshore corporation, keeping offshore transactions confidential. Offshorecorporation.com can help obtain an offshore bank account and corporate offshore bank account.

Having an offshore corporation with an offshore bank account that is titled to the offshore corporation, your assets are taken off the radar screen. Establishing an offshore company in Nevis, and an offshore bank account in Bermuda or Switzerland, will help provide you with maximum privacy. The Caribbean Island of Nevis located East of Puerto Rico and Southeast of Florida. Information on the owner of a Nevis LLC (Limited Liability Company) is not public record and is therefore confidential.

Looking for protection from divorce, lawsuits, and legal but controversial enterprises, such as online pharmacies or cyber casinos? Banking provides the privacy and protection required for these types of situations. Need protection of financial resources or privacy of ownership? Then come to the banking specialists at Offshorecorporation.com

Offshore is more affordably priced than one may think. Establishing an offshore corporation plus an offshore bank account at the same time is the best economical route. Many jurisdictions can be as much as $5000 to $8000. However, a Nevis LLC is valued at only $1495 and an offshore bank account is only $350 depending on the jurisdiction. That is a total of only $1845 US to own an offshore corporation and offshore bank account. You can call 800-830-1055 or 661-255-3303 for more information.

INTERNATIONAL CELL PHONE COVERAGE

www.cellularexpressphonerentals.com

Headquarters:
1430 Broadway
New York, NY 10018
Toll Free: 1-800-320-7776

For any questions, please e-mail us at info@cellularexpressphonerentals.com.

Cell Phone Rentals For Domestic & International Wireless Communication

Since 1995 Cellular Express Phone Rentals, Inc. has been an industry leader in short-term international **cell phone rentals** for the global business and leisure traveler. We are continuously upgrading our equipment fleet of GSM cellular phones, satellite phones and other international cellular phone systems, as well as expanding our global reach in order to provide the latest state-of-the-art cell phone rental solutions for people in need of immediate seamless worldwide mobile communication.

Cellular Phone Rentals You Can Count On

Whether you are traveling to Europe, Asia, Africa, Middle East, North, Central or South America, Mexico, Caribbean or Australia, we will have cellular phone rentals to suit your needs. With international cellular service virtually spanning the globe, you can be rest assured we cover the areas, which you will be traveling to.

Cell Phone Rentals with Worldwide Coverage

Our extensive client base of fortune 500 companies, government agencies as well as thousands of individual world travelers can attest to our reputation of unsurpassed quality, dependability and expertise in the field of international cell phone rentals and satellite phone rentals.

www.worldcell.com

WorldCell is the leader in providing global wireless service solutions to Government, Enterprise, Industry as well as the international traveler. WorldCell provides solutions through a number of technologies, ranging from an international cell phone or satellite phone to the Roaming Service Bureau platform that mediates and translates different wireless standards for wireless network operators.

WorldCell leverages its GSM operating license, operating platforms and global partnerships to deliver the most advanced and seamless wireless communications services available today on a global scale....no matter where in the world you need to be connected.

Global Reach – No matter where you are, WorldCell has a solution for where you live and where you travel.

Local Commitment - WorldCell can provide the best in local service just about anywhere in the world. From Washington DC to Frankfurt to Bahrain, we provide the solutions only the expert in local communication infrastructure can provide.

A World of Experience - WorldCell has been serving the global marketplace and providing solutions to the enterprise and industry in the ever changing wireless landscape since 1996. Backed by a team of global telecom experts, best in breed partners and an unparalled technology platform, WorldCell has the proven experience, and solutions to meet your wireless communications needs – globally.

You may contact WorldCell™ in the following ways:

Toll Free: 1.888.967.5323
Worldwide : 1.301.960.0060
Customer & Technical Care Direct Line : 1.301.960.0078
Government Team Direct Line : 1.800.436.0244

info@worldcell.com
www.worldcell.com

CORPORATE OFFICE WorldCell:

801 Roeder Road
Suite 800
Silver Spring, Maryland 20910/Phone: 888.967.5323/301.960.0060

ICELAND OFFICE WorldCell

IMC Island ehf.
20 Armuli
Reykjavik 108
Iceland/Phone: 354.588.8540

BRAZIL OFFICE WorldCell

International Mobile Communications Brazil/Phone: 55.21.3970.1093

GERMANY OFFICE WorldCell

Phone: 49.173.294.4917

CASTING DIRECTORS - TV

Steve Brooksbank
c/o CSA
606 N. Larchmont Blvd. Suite 4-b
Los Angeles, California 90004
UNITED STATES
(818) 777-6572

Series
Threat Matrix Touchstone TV ABC
Line of Fire Touchstone/Dreamworks ABC

April Webster
April Webster Casting
513 Wilshire Blvd. Suite 196
Santa Monica, California 90401
UNITED STATES
(818) 526-4242

Series
Alias Touchstone ABC

Deedee Bradley
6767 Forest Lawn Drive Suite #100
Los Angeles, California 90068
UNITED STATES
818-977-8956

Series
Smallville (LA) Warner Bros. WB

Georgianne Walken
c/o Csa
145 W. 28th Street Suite 12F
New York, New York 10001
UNITED STATES
(212) 877-4025

Series
The Sopranos Brillstein-Grey HBO
Lynn Kressel

Robin Lippin
CSA
606 N. Larchmont Blvd. Suite 4B
Los Angeles, California 90004
UNITED STATES
(818) 553-5804

Series
Happy Family NBC Prods. NBC
Lizzie McGuire
State of Grace

Meg Liberman
Liberman/Patton Casting
6464 Sunset Blvd. Suite 707
Los Angeles, California 90028
UNITED STATES
(323) 462-9175

Series
Boomtown Dreamworks NBC
King of Queens Columbia/Tri-Star CBS
Las Vegas Dreamworks/NBC
Ground for Life Casrsey-Werner-Mandebach FOX

Holly Powell
Powell / Melcher Casting
5555 Melrose Ave. Spec. Effects Bldg.
2nd Floor
Los Angeles, California 90038
UNITED STATES
(323) 956-4260

Series
One on One Greenblatt-Janollari UPN
The Opposite of Sex Warner Bros UPN

Debi Manwiller
Pagano/Manwiller Casting C/o Csa
606 N. Larchmont Blvd. Suite 4b
Los Angeles, California 90004
UNITED STATES
(323) 463-1925

Series
24 Imagine TV Fox

Nikki Valko
Valko/Miller Casting
3500 W. Olive Ave. Suite 780
Burbank, California 91505
UNITED STATES
(818) 953-7743

Series
The Brotherhood of Poland, NH 20th CBS
Malcolm in the Middle Regency Fox

Gillian O Neill
C/o Csa
606 N. Larchmont Blvd. Suite 4-b
Los Angeles, California 90004
UNITED STATES
(323) 463-1925

Series
Game Over C-W-M UPN
Judging Amy 20th Century Fox CBS
Dharma & Greg 20th Century Fox ABC
Celebrity Death Match MTV2

Pilot
Gramercy Park Warner Bros. ABC
Joint Custody Warner Bros. WB
Lennie Rose Touchstone UPN
20 Things to Do Before You're 30 Touchstone UPN

Greg Orson
Greg Orson Casting
6464 W. Sunset Blvd Ste 970
Los Angeles, California 90028
UNITED STATES
(323) 469-6464

Feature		
Cursed	Dimension Films	Wes Craven
Movie of the Week		
Romy & Michele: In The Beginning	Touchstone Television	ABC
Series		
Reba	20th Century Fox	WBN
Run of the House	Warner Bros. Television	WBN
Shacking Up	Regency Television	WBN
Miss Match	NBC Studios	NBC
Dawson's Creek	Sony Television	WBN
Glory Days	Miramax Television	WBN
Grosse Pointe	Artist Television Group	WBN
Son of the Beach	Fox Television Studios	FX
Loose Cannon	Warner Bros. Television	WBN
The Secret Diary of Desmond Pfeiffer	Paramount Television	UPN
Hitz	Paramount Television	UPN
Claude's Crib	Paramount Television	USA
Pilot		
Three	Paramount Television	CBS
Rocky Point	Warner Bros. Television	WBN
Reba	20th Century Fox	WBN
Shacking Up	Regency Television	WBN
Weekends	NBC Studios	NBC
Mr. Ed	20th Century Fox	FBC
Sixteen to Life	20th Century Fox	WBN
Exit 9	Warner Bros. Television	WBN
Trash	Regency Television	WBN

Feature

Glory Days	Miramax Television	WBN
In My Opinion	Turner Films	WBN
Waydowntown	Brad Grey Television	WBN
Grosse Pointe	Artist Television Group	WBN
Son of the Beach	Fox Television Studios	FX
Electra Woman & Dyna Girl	Beverly Actors Company	WBN
Untitled Harold Ramis Project	Brad Grey Television	WBN
Fresh Meat	Michigan J. Productions	WBN
Minor Threat	Michigan J. Productions	WBN
The Secret Diary of Desmond Pfeiffer	Paramount Television	UPN
Claude's Crib	Paramount Television	USA

Fran F. Bascom
Fran F. Bascom & Assoc.
3400 Riverside Dr Suite 765
Burbank, California 91505
UNITED STATES
(818) 972-8339

Series
Days of Our Lives Columbia Pictures TV NBC

Carol Kelsay
Carol G. Kelsay Casting, Inc.
225 W. 2nd Avenue
Vancouver, B.C., Non-US V5y 1c7
UNITED STATES
(604) 263-1105

Series
Stargate MGM Sci-Fi

Debby Romano
Romano/Benner Casting C/o Csa
606 N. Larchmont Blvd. Suite 4-b
Los Angeles, California 90004
UNITED STATES
(323) 463-1925

Series
Scrubs Touchstone NBC
What I Like About You Warner Bros. WB

Mark Saks
CBS
51 W. 52nd Street 20th Floor
New York, , New York 10019
UNITED STATES
(213) 534-3060

Series
Tarzan Warner Bros. TV WB
Hack Big Ticket/CBS Prods. CBS

Lynne Carrow
Lynne Carrow
Carrow Brouse Casting 23 West Pender St.
Ste. 405
Vancouver, British Columbia V6B 1R3
CANADA
604-639-1804

Series
Da Vinci's Inquest Da Vinci Prods. CBC

Leslee Dennis
c/o CSA
606 N. Larchmont Blvd. Suite 4B
Los Angeles, California 90004
UNITED STATES
(323) 463-1925

Series
Charmed Spelling Ent. WB

Meg Simon
Warner Bros. TV
1325 6th Avenue
New York, New York 10019
UNITED STATES
(212) 636-5145

Series
All About The Anderson (NY) Warner Bros. TV WB

Holly Powell
Powell / Melcher Casting
5555 Melrose Ave. Spec. Effects Bldg.
2nd Floor
Los Angeles, California 90038
UNITED STATES
(323) 956-4260

Series
One on One Greenblatt-Janollari UPN
The Opposite of Sex Warner Bros. UPN

Gary M Zuckerbrod
C/o Csa
606 N. Larchmont Blvd. Suite 4-b
Los Angeles, California 90004
UNITED STATES
(323) 463-1925

Series
Without A Trace Warner Bros/CBS Prods. CBS

Liz Marx
C/o Csa
606 N. Larchmont Blvd. Suite 4-b
Los Angeles, California 90004
UNITED STATES
(323) 463-1925

Series
Like Family Warner Bros. TV WB

Coreen Mayrs
Coreen Mayrs Casting, Inc
555 Brooksbank Ave. #120 Bldg. 9
North Vancouver, B.C., B.C. V7j 3s5
Canada
(604) 983-5675

Series
Smallville (Vancouver) Warner Bros. WB

Junie Lowry Johnson
20th Century Fox
10201 W. Pico Blvd. Bochco #232
Los Angeles, California 90035
UNITED STATES
(310) 369-1296

Series
Deadwood Roscoe Prods. HBO
Six Feet Under Greenblatt-Janollari HBO
Tremors Studios USA Sci-Fi
NYPD Blue Bochco Prods. ABC
Enterprise Paramount TV UPN

Tracy Lilienfield
Tracy Lilienfield Casting
4024 Radford Avenue Sound Shop 2nd floor
Studio City, California 91604
UNITED STATES
(818) 655-5652

Series
Will & Grace NBC Studios NBC
The Stones Warner Bros. TV CBS
Good Morning, Miami Warner Bros. TV NBC

Jeff Greenberg
Los Angeles, California
UNITED STATES

Series
Frasier Grubb Street Prods. NBC
I'm All Relative Paramount/Touchstone/Storyline ABC
I'm With Her Tollin-Robbins/Warner Bros. ABC
According To Jim Touchstone/ABC ABC

Sheila Jaffe
Walken & Jaffe Casting
Los Angeles, California
UNITED STATES
(323) 337-0595

Series
The Sopranos Brillstein-Grey HBO

Reuben Cannon
Reuben Cannon & Associates
5225 Wilshire Blvd Suite 526
Los Angeles, California 90036
UNITED STATES
(323) 939-3190

Series
The Parkers Big Ticket TV UPN
Half and Half CBS Prods. UPN

Lisa Miller Katz
C/o Csa
606 N. Larchmont Blvd. Suite 4b
Los Angeles, California 90004
UNITED STATES
(818) 655-5797

Series
Everybody Loves Raymond Worldwide Pants/Hip CBS

Vicki Rosenberg
20th Century Fox
10201 W. Pico Blvd. Bldg. 80 Room 10
Los Angeles, California 90035
UNITED STATES
410) 578-3847

Series
Joan of Arcadia 20th Century Fox CBS
Judging Amy 20th Century Fox CBS

Suzanne Ryan
Universal Television
Pier 62 Room 304 W. 23rd St. & Hudson River
New York, New York 10011
UNITED STATES
(212) 414-2941

Series
Law & Order Universal NBC

Patrick J Rush
Warner Bros.
300 Television Plaza Bldg 140 Room 134
Burbank, California 91505
UNITED STATES
(818) 954-5126

Series
The O.C. Warner Bros/Hypnotic Fox
Everwood Everwood Utah, Inc. WB

Paul Weber
MGM
10250 Constellation Blvd. 2nd Floor
Los Angeles, California 90067-6247
UNITED STATES

Series
Dead Like Me (L.A.) MGM Showtime
Stargate MGM Sci-Fi
She Spies NBC Prods. NBC/Syn
Jeremiah MGM Showtime

Mark Teschner
General Hospital
4151 Prospect Avenue
Hollywood, California 90027
UNITED STATES
(323) 671-5542

Series
General Hospital ABC Productions ABC

Mary Jo Slater
Slater-Brooksbank C/o USA Television
100 Universal City Plaza Bldg. 1440 14th Floor
Universal City, California 91608
UNITED STATES
(818) 777-6572

Series

Line of Fire	Touchstone/Dreamworks	ABC
Peacemakers	Michael R. Joyce Prods.	USA
Threat Matrix	Touchstone TV	ABC

Jacklynn Briskey
C. B. S. Studio Center
4024 Radford Admin. Bldg. Suite 280
Studio City, California 91604
UNITED STATES
(818) 655-5601

Series

Passions	NBC Studios	NBC

Rick Millikan
20th Century Fox
10201 W. Pico Blvd Bldg 71
Los Angeles, California 90035
UNITED STATES
310/369-4447

Movie of the Week		
CHRISTMAS ON DIVISION STREET		COLUMBIA/CBS
I STILL DREAM OF JEANNIE		COLUMBIA/NBC
REVOLVER		COLUMBIA/NBC
PRIME TARGET		MGM/NBC
Short Feature		
KANGAROO COURT	ACADEMY AWARD NOMINEE (BEST SHORT)	LAVA ENT./SEAN ASTIN PRODS.
Series		
CSI:MIAMI	EPISODES 1-12	CBSP/BRUCKHEIMER/CBS
THE X-FILES	ALL 201 EPISODES	20TH CENTURY FOX/FBC
SABRINA THE TEENAGE WITCH	7 SEASONS	VIACOM/ABC/WB
MARRIED WITH CHILDREN	LAST 2 SEASONS	COLUMBIA/FBC
RAISING CAINES		TRI-STAR/NBC
ROUTE 66		COLUMBIA/NBC
IN THE HEAT OF THE NIGHT		MGMTV/NBC
THE YOUNG RIDERS		MGMTV/ABC
The Guardian	Columbia	CBS
QUINTUPLETS		20TH/IMAGINETV/FBC
Pilot		
COURTSHIP OF EDDIE'S FATHER(2003)		WBTV/PARIAH/WB
STRANGE FREQUENCY		VIACOM/VH1
LIFE'S TOO SHORT		20TH CENTURY FOX/WB
THE HANLEY'S		BRILLSTEIN/GREY/ABC
SKIP CHASERS		20TH CENTURY FOX/CBS
BEDTIME		VIACOM/SHOWTIME
KINDRED: THE EMBRACED		SPELLING/FBC
ROUTE 66		COLUMBIA-PROPAGANDA/NBC
NEW YEAR		COLUMBIA/ABC
AGAINST THE MOB		MGM/NBC
SWEDEN, OHIO		WBTV/FBC
SECOND NATURE		20TH/IMAGINE/FBC

Additional

DIRECTOR OF TALENT & CASTING	COLUMBIA TV	1990-92
DIRECTOR OF TALENT & CASTING	MGM/UA TV	1987-90

John Buchan
John Buchan Casting, Inc.
499 Sackville St.
Toronto, Ont., Alabama M4x 1t6
UNITED STATES
(416) 927-9363

Series
The Strip (Toronto) Saftsbury Sky Films YTV

Peggy Kennedy
Pagano / Manwiller Casting
6250 Canoga Ave.
Woodland Hills, California 91367
UNITED STATES
(818) 615-2146

Series
24 Imagine/20th Century Fox

Vicki Huff
Victoria Huff & Assoc.
5700 Wilshire Blvd. 5 North Suite 5038
Los Angeles, California 90036
UNITED STATES
(323) 634-1260

Movie of the Week

20,000 Leagues Uner The Sea	ABC
The Substitute Wife	NBC
The Absolute Truth	CBS
Lily In Winter	USA
Falsely Accused	CBS
Dead Air	USA
Sworn To Silence	ABC
The Comrades Of Summer	HBO
Echoes In The Darkness	CBS

Series

Fat Actress	Showtime
Seventh Heaven	WB
Charmed	WB
Tales From The Crypt	HBO
Beverly Hills 90210	FOX
Wiseguy	CBS
Sliders	Fox
SeaQuest DSV	NBC
2000 Malibu Road	CBS
Extreme	ABC

Brett Benner
Romano/Benner Casting C/o Csa
606 N. Larchmont Blvd. Suite 4-b
Los Angeles, California 90004
UNITED STATES
(818) 954-5470

Series
Scrubs Touchstone NBC
What I Like About You Warner Bros. WB

Mark Saks
CBS
51 W. 52nd Street 20th Floor
New York, , New York 10019
UNITED STATES
(213) 534-3060

Series
Tarzan Warner Bros. TV WB
Hack Big Ticket/CBS Prods. CBS

Roger Mussenden
Roger Mussenden Casting
10536 Culver Blvd. Suite C-1
Culver City, California 90232
UNITED STATES
(310) 559-9522

Series
American Family Greenblatt-Janollari PBS

John Papsidera
Automatic Sweat, Inc.
5541 W. Washington Blvd.
Los Angeles, California 90016
UNITED STATES
(323) 934-5141

Series
Carnivale Carney Prods. HBO

Rob Decina
Guiding Light
524 West 57th Street Suite 5320
New York, New York 10019
UNITED STATES
www.artofauditioning.com
(212) 986-5330

Series
Guiding Light Procter & Gamble CBS

Sharon Bialy
Bialy / Thomas Casting
16000 Ventura Blvd. #301
Encino, California 91436
UNITED STATES
(818) 386-1894

Series
Skin Warner Bros/Bruckheimer Fox

Megan Mcconnell
Gilmore/McConnell Casting
20th Century Fox 10201 W. Pico Blvd.
LA, California 90035
UNITED STATES
310-369-3755

Movie of the Week		
War Stories	NBC Prods.	NBC
Series		
The Practice	David E. Kelley Productions for ABC	ABC
Dragnet	Universal	ABC
Felicity	Imagine TV/Touchstone	WB
Snoops	David E. Kelley Prods/20th TV	ABC

Eric Dawson
Ulrich/Dawson/Kritzer
4705 Laurel Canyon Blvd. Suite 301
Valley Village, California 91607
UNITED STATES
(818) 623-1818

Feature		
Wheelmen	Stretch Pix Prods.	Dir: Dirk Hagen 2001
Cold Hearts	Winner: Best Film - Atlantic City Film Festival	Dir: Robert Masciantonio 2000
Airborne	Icon Pictures/Warner Bros.	Dir: Rob Bowman 1995-1996
Monolith	Tucker and Flynn Prods.	Dir: John Eyres
Spies on Ice	Vestron Pictures	Dir: Anthony Thomas 1990
Movie of the Week		
Homeland Security	Paramount TV	NBC MOW/Pilot 2003
Battlestar Gallactica	Aces and Eights/USA Cable	USA Miniseries 2003
Dark Angel	American Zoetrope/Big Ticket	Fox MOW 1996
She Cried No	Steve White Ent.	NBC MOW 1996
Gramps	Viacom Prod.	NBC MOW 1995
The Adventures of Captain Zoom	Universal/Telvan Prod.	Bravo MOW 1995
Perry Mason	Viacom Prod.	12 NBC MOWs 1990-1994
Ray Alexander	Viacom Prod.	NBC MOW 1994
This Gun for Hire	BBK Productions	USA MOW 1990
Series		
The Lyon's Den	20th Century Fox/Brad Grey TV	NBC Series 2003
Nip/Tuck	Shephard/Robin Productions	F/X Pilot and Series 2003
1-800-MISSING	Lions Gate TV	Lifetime Series 2003
CSI: Crime Scene Investigation	Alliance Atlantis & CBS Prods.	CBS Series 2001-2003
Push, Nevada	LivePlanet & Touchstone TV	ABC Series 2002
The Agency	Radiant Prods, CBS Prods, Studios USA	CBS Series 2001-2003
MD's	Touchstone TV	ABC Series 2002
The Dead Zone	Lions Gate TV/Paramount Int.	USA Series 2002-2003
Felicity	Imagine TV & Touchstone TV	WBN Series 2001-2002
Odyssey V	Columbia/Tri-Star TV	Showtime 2-hr Pilot & Series 2001-2002
The American Embassy	20th Century Fox/Jersey TV	Fox Series 2001-2002
The Chris Isaak Show	Viacom Prod.	Showtime Series 2001-2002
Dark Angel	20th Century Fox	Fox Pilot & Series 1999-2002
Any Day Now	Paid Out Dues Prods.	Lifetime Series 1998-2002
The Tick	Columbia/Tri-Star TV	Fox Series 2000-2001
Popular	Touchstone TV	WBN Pilot & Series 1999-2001
Early Edition	Columbia/Tri-Star TV	CBS Series 1997-2000
Brutally Normal	Touchstone TV	WBN Pilot and Series 1999-2000
Ryan Caulfield: Year One	Regency	Fox Series 1999
To Have & To Hold	Greenblatt/Janollari & CBS Prods.	CBS Series 1998
Rude Awakening	Mandalay	Showtime Pilot & Series 1998-1999
The Gregory Hines Show	Columbia/Tri-Star TV	CBS Pilot & Series 1997-1998
Timecop	Universal TV	ABC Pilot & Series 1997
Dark Skies	Columbia/Tri-Star TV	NBC Series 1996-1997
Spy Game	Universal TV/Warner Bros. TV	ABC Pilot & Series 1996-1997
Moloney	Columbia/Tri-Star TV	CBS Series 1996-1997
Pensacola	3 Arts Ent.	Synd. Pilot & Series 1997
Strange Luck	New World Ent.	Fox Pilot & Series 1995-1996
Due South	Alliance Communications	CBS Series 1995-1996
Deadly Games	Viacom Prod.	UPN Pilot & Series 1995-1996

The Byrds of Paradise	Steven Bocho Prods.	ABC Pilot & Series 1994
Melrose Place	Aaron Spelling Productions	Fox Series 1992-1993
Ned Blessing...Return to Plum Creek	Hearst Ent.	CBS Pilot & Series 1993
Eerie, Indiana	Hearst Ent.	NBC Pilot & Series 1991-1992
The Round Table	Aaron Spelling Productions	NBC Pilot & Series 1992
Pilot		
Hope and Faith	Touchstone TV	ABC Pilot 2003
Crazy Love	Touchstone TV	CBS Pilot 2003
Expert Witness	20th Century Fox/The Firm	CBS Pilot 2003
The Circle aka Alaska	Touchstone TV	ABC Pilot 2003
CSI: Miami	Alliance Atlantis & CBS Prods.	CBS Pilot 2002
Aces	Shephard/Robin, Regency TV, Warner Bros. TV	WBN Pilot 2002
Untitled Anne Rice Project	Imagine TV & 20th Century Fox	NBC Pilot 2001
Sam's Circus	Columbia/Tri-Star TV	CBS Pilot 2001
The Chronicle	Greenblatt/Janollari & NBC Studios	NBC Pilot 2000
Untitled Ganz/Mangell Project	Imagine TV	ABC Pilot 2000
Day One	Regency/Granada	WBN Pilot 2000
Sagamore	Touchstone TV	WBN Pilot 1999
Partners	Columbia/Tri-Star TV	CBS Pilot 1999
Glory, Glory	Aaron Spelling	CBS Pilot 1998
Daybreak	20th Century Fox	Fox Pilot 1998
Anne Rice's Rag & Bone	Columbia/Tri-Star TV	CBS Pilot 1997
The Dave Chappelle Project	Disney	Fox Pilot 1997
Dogs	Disney	Touchstone Pilot 1997
The Gail O'Grady Project	Columbia/Tri-Star TV	Fox Pilot 1996
Hell	20th Century Fox	Fox Pilot 1995
Fast Company	Viacom Prod.	NBC 2-hour Pilot 1995
Close to Home	Witt/Thomas Prods.	NBC Pilot 1994
Bringing Up Jack	a.k.a. Productions	ABC Pilot 1994
Arly Hanks	Hearst Ent.	CBS Pilot 1993
Green Dolphin Street	Aaron Spelling Productions	Fox Pilot 1993
Little Man Tate	Stone/Foster/Frank Prods.	NBC Pilot 1992
The Hannigans	Viacom Prod.	Fox Pilot 1992

Diane Heery
Heery Casting, LLC
230 N. 2nd Street Suite 1-a
Philadelphia, Pennsylvania 19106
UNITED STATES
www.heerycasting.com
(215) 238-9240

Feature		
The Woodsman	location principals and extras	2002
Jersey Girl	location principals and extras	2002
SIGNS	location principals and extras	2001
National Treasure	Location principals and extras	2003
Shadowboxer	location principals and extras	2004
The Final Patient	Location principals	2004
Invincible	location principals and extras	2005
Series		
Hack	location principals and extras	CBS
Cold Case	location principals and extras	CBS
Jamie Kennedy Experiment	location principals and extras	WB

Feature

Pilot

The Block Series regulars The N

Joey Paul Jensen
Hollywood Center Studios
1040 N. Las Palmas Bldg. 33
Hollywood, California 90038
UNITED STATES
(323) 860-8989

Series
That's So Raven Disney TV Disney Channel
Phil Of the Future 2121 Prods. Disney Channel
Pajama Rama Sweet Lorraine Prods. WAM

Alex Wald
C/o Csa
606 N. Larchmont Blvd. Suite 4-b
Los Angeles, California 90004
UNITED STATES
(323) 463-1925

Series
One Tree Hill Tollin-Robbins WB

Kevin Scott
CBS
4024 Radford Ave/ Bldg.5 Room 105
Studio City, California 91604
UNITED STATES
(818) 655-6379

Series
All About the Andersons (LA) Warner Bros. TV WB
All of Us Overbrook/Warner Bros. UPN
My Wife and Kids Touchstone ABC

Janet Gilmore
Gilmore/McConnell Casting
1800 Stewart St., 1st floor
Santa Monica, California 90404
UNITED STATES
(310) 579-5180

Series
The Practice 20th Century TV ABC

Megan Branman
C/o Csa
606 N. Larchmont Blvd. Suite 4-b
Los Angeles, California 90004
UNITED STATES
(323) 463-1925

Series
LA Dragnet Universal/Studios USA ABC

Geraldine Leder
C/o Csa
606 N. Larchmont Blvd. Suite 4-b
Los Angeles, California 90004
UNITED STATES
(323) 463-1925

Series
Cracking Up 20th Century/Brad Grey TV Fox
Oliver Beene Dreamworks/20th Century 20th Fox

Jason La Padura
La Padura & Hart Casting
100 Universal City Plaza Mt 6149
Universal City, California 91608
UNITED STATES
(818) 733-4735

Series
American Dreams Universal/NBC Prods. NBC
Crossing Jordan NBC Studios NBC

Steve Brooksbank
C/o CSA
606 N. Larchmont Blvd. Suite 4-b
Los Angeles, California 90004
UNITED STATES
(818) 777-6572

Series
Threat Matrix Touchstone TV ABC
Line of Fire Touchstone/Dreamworks ABC

Gilda Stratton
C/o Csa
606 N Larchmont Blvd. Suite 4-b
Los Angeles, California 90004
UNITED STATES
(323) 463-1925

Series
A Minute With Stan Hooper 20th/Parmamount Fox

Tracy Lilienfield
Tracy Lilienfield Casting
4024 Radford Avenue Sound Shop 2nd floor
Studio City, California 91604
UNITED STATES
(818) 655-5652

Series
Will & Grace NBC Studios NBC
The Stones Warner Bros. TV CBS
Good Morning Miami Warner Bros. TV NBC

Geraldine Leder
C/o Csa
606 N. Larchmont Blvd. Suite 4-b
Los Angeles, California 90004
UNITED STATES
(323) 463-1925

Series
Cracking Up 20th/Brad Grey TV Fox
Oliver Beene Dreamworks/20th Fox

Dava Waite
C/o Csa
606 N. Larchmont Blvd. Suite 4-b
Los Angeles, California 90004
UNITED STATES
(323) 463-1925

Series
Yes, Dear 20th TV CBS

Wendy O Brien
Automatic Sweat
5541 W/ WASHINGTON BLVD.
Los Angeles, California 90016
UNITED STATES
(323) 934-5141

Series
Carnivale Carney Productions HBO

Jeff Greenberg
Los Angeles, California
UNITED STATES

Series
Frasier Grubb Street Prods. NBC
I'm With Her Tollin-Robbins/Warner Bros.
It's All Relative Paramount/Touchstone/Storyline ABC
According to Jim Touchstone/ABC ABC

John Levey
Warner Bros.
300 S Television Plaza Bldg 140 Room 138
Burbank, California 91505
UNITED STATES
(818) 954-4080

Series
ER Amblin/Warner Bros. NBC

Stuart Aikins
Aikins/Cossey Casting
100 W Pender St #401
Vancouver, British Columbia V6B 1R8
CANADA
(604) 739-4612

Series

Andromeda	Fireworks/Tribune	SYN

Leslie Litt
C/o Csa
606 N. Larchmont Boulevard Suite 4-b
Los Angeles, California 90004
UNITED STATES
(323) 463-1925

Series

Friends	Warner Bros. TV	NBC

Cami Patton
Liberman/Patton Casting
6464 Sunset Blvd. Suite 707
Los Angeles, California 90028
UNITED STATES
(323) 462-9175

Feature		
Eurotrip	Montecito/Dreamworks	
My Big Fat Greek Wedding	Playtone/Gold Circle	
Series		
Boomtown	Dreamworks	NBC
Las Vegas	Dreamworks/NBC Studios	NBC
Grounded for Life	Carsey-Werner-Mandebach	Fox
King of Queens	Sony/CBS Productions	CBS
Out of Order	Showtime	Showtime
Pasadena	BGE/Sony	Fox
Other		
Band of Brothers	Playtone	HBO
Anne Frank		ABC

Stephanie Gorin
Stephanie Gorin Casting
62 Ellerbeck St.
Toronto, Ont., Ontario M4k 2v1
CANADA
(416) 778-6916

Feature		
Cavedweller	Showtime	Lisa Cholodenko
Godsend	Lions Gate	Nick Hamm
Chicago (Co-casting Toronto)	Miramax	Rob Marshall
Movie of the Week		
Untitled Britney Spears Project	ABC/Lions Gate	Bobby Roth
Cheetah Girls	Disney Channel	Oz Scott
Deacons For Defense	Showtime	Bill Duke
The Winning Season	TNT	John Kent Harrison
Comfort and Joy	Paramount/Lifetime	Maggie Greenwald
DC 911	Paramount/Showtime	Brian Tranchard Smith
Out of the Ashes	Showtime	Joe Sargeant
Crown Heights	Showtime	Jeremy Kagan

Feature

Jett Jackson (The Movie)	Disney Channel	Shawn Levy
Whitewash	Paramount/Viacom	Tony Bill
Feast of All Saints	Showtime	Peter Medak
Short Feature		
The Truth About Head	Cannes Film Festival Winner	Dale Heslip
Series		
Sue Thomas, F.B. Eye (Toronto)	Pax	Various directors
1 800 Missing (Toronto)	Lions Gate/Lifetime	Various directors
Pilot		
The Coven	Lions Gate	Lifetime
Jake 2.0	Viacom	Rob Lieberman
Theater		
Mamma Mia (3 companies)	Littlestar Productions	Phyllida Lloyd
The Lion King (Canada)	Disney Theatrical	Julie Taymor

Bonnie Zane
Zane/Pillsbury Casting
585 N. Larchmont Blvd.
Los Angeles, California 90004
UNITED STATES
(323) 769-9191

Series

Miss Match	20TH/Imagine	NBC
Hope & Faith	Industry/Touchstone	ABC

Meg Liberman
Liberman/Patton Casting
6464 Sunset Blvd. Suite 707
Los Angeles, California 90028
UNITED STATES
(323) 462-9175

Series

Boomtown	Dreamworks	NBC
King of Queens	Columbia/Tri-Star	CBS
Las Vegas	Dreamworks/NBC Studios	NBC
Grounded for Life	Carsey-Werner-Mandebach	FOX

Alexa I Fogel
Beech Hill Films
443 Greenwich Street Suite 5a
New York, New York 10013
UNITED STATES
(212) 226-3331

Series

Baseball Wives	Levinson Productions	NBC

Felicia Fasano
c/o CSA
606 N. Larchmont Blvd. Suite 4-B
Los Angeles, California 90004
UNITED STATES
(818) 260-2655

Series
The Handler Viacom CBS

Bernard Telsey
Bernard Telsey Casting, Inc.
145 W. 28th Street Suite 12f
New York, New York 10001
UNITED STATES
(212) 868-1260

Series
Whoopi Carsey-Werner-Mandebach NBC

Patrick Baca
Casting Society of America
606 N. Larchmont Blvd. Suite 4B
Los Angeles, California 90004
UNITED STATES
(818) 528-2080

Feature		
Tweek City	Indie Feature	Stars Giuseppe Andrews
Halloween: Resurrection	Miramax/Dimension	Stars Busta Rhymes
Defying Gravity	Jour De Fete	Stars Nicklaus Lang
Just Can't Get Enough	Regent Ent.	Stars JP Pitoc
Tryst	MCEG/Sterling	Stars Louise Fletcher
Hard Pill	Stoebner/Baumgardner Co.	Stars Jonathan Slavin, Jack Plotnick
Movie of the Week		
L.A. Law: The Movie	NBC/20th	Stars Bruce Davison
Gentle Ben 1 & 2	Hallmark	Stars Dean Cain
The Christmas Secret	CBS	Stars Richard Thomas
Short Feature		
Size 'em Up	Girls on Film	Stars Julie Brown
Boychick	Boxer Shorts 2003	Stars Ben Lange
Breaking Up Really Sucks	Girls on Film	Stars Alexandra Paul
L.T.R.	Boys Life: Four Play	Stars Cole Williams
Jason Scott	Amalgamated Films	Stars French Stewart
Series		
Stripmall	Comedy Central	Starring Julie Brown
Foto-Novelas 2	Stars Robert Beltran	PBS
Paul Frank's Julius & Friends	Animated Web Series	www.paulfrank.com
Mad TV	Fox	Series Regular Search
Pilot		
Alligator Point	NBC/Paramount	Stars Park Overall, Jaime Pressly
Daydreams	Disney Channel	Pro: Matt Freeman
Hell House	MTV	w/Glicksman-Orenstein
The Lot	AMC	w/Joel Thurm
Reunited	Columbia/Tri-Star	w/Vicki Rosenberg

Maggie Day	Lifetime	w/Judith Weiner
Dads	Columbia/Tri-Star	Starring Rue McClanahan
Marry My Mom	CBS/Columbia/Tri-Star	w/Vicki Rosenberg
Movie Surfers	Disney Channel	Stars Matt Botuchis
Stage		
Three Grooms & A Bride	Coast Playhouse	Starring Anne DeSalvo
Parental Discretion	Coast Playhouse	Starring John Aniston
End of the World Party	Celebration Theatre	Starring Jim J. Bullock
Troubled Waters	Ventura Court	Starring Cynthia Gibb
Club Hell	Celebration Theatre	DramaLogue Award - Casting Direction
CyberQueer	Celebration Theatre	Starring Tom Paris
Areas of Casting		
Features	Independent Film	Indies & Shorts
Movies of the Week & MOWs	Episodics & Pilots	Reality
Theatre & Stage	Regional	Equity Waiver
Latino Latin Hispanic Themes	Gay Themes	

Tammara Billik
Tammara Billik Casting
13547 Ventura Blvd. #688
Sherman Oaks, California 91423
UNITED STATES
(818) 623-1631

Feature		
The Peacemakers	USA Studios	USA
The Ellen Show	ATG	CBS
Series		
The Help	Warner Bros.	WB
Ellen	Disney	ABC
Married...with Children	Columbia TriStar	FBC
Unhappily Ever After	Disney	WB
Three Sisters	NBCP	NBC
Honey, I Shrunk the Kids	Buena Vista	SYND

Gayle Pillsbury
Zane/Pillsbury Casting
585 N. Larchmont Blvd.
Los Angeles, California 90004
UNITED STATES
(323) 769-9191

Series		
The George Lopez Show	Warner Bros. TV	ABC
Drew Carey Show	Warner Bros.	ABC

Judy Blye Wilson
ABC All My Children
320 W 66th
New York, New York 10023
UNITED STATES
(212) 456-0802

Series

All My Children ABC Productions ABC

Linda Lowy
Linda Lowy Casting
4151 Prospect Ave. Cottages, Room 105
Los Angeles, California 90027
UNITED STATES
(323) 671-5438

Series
Karen Sisco Universal/Jersey TV ABC

Deborah Barylski
CBS Studio Center
4024 Radford Ave.
Bldg. 2, Room 100-102A
Studio City, CA,

Pilot
Flying Sauces
Reach Out and Teach Someone
Look Who's Not Talking
Yule Better Watch Out
Up Your Alley
For Whom the Belch Tolls
Forever Jung
Jill's Birthday
What About Bob?
Baby, It's Cold Outside
Unchained Malady
Birds of a Feather Flock to Taylor
A Battle of Wheels
Luck Be a Taylor Tonight
Al's Fair in Love and War
Stereo-Typical
Read My Hips
Rites & Wrongs of Passage
Overactive Glance
Groin Pains
Heavy Meddle
The Haunting of Taylor House
Roomie for Improvement
May the Best Man Win
Where There's A Will, There's A Way
Let's Did Lunch
Abandoned Family
I'm Scheming of a White Christmas
Bell Bottom Blues
Howard's End

Love is a Many Splintered Thing
Dances with Tools
You're Driving Me Crazy, You're Driving Me Nuts
Bye, Bye Birdie
Karate or Not, Here I Come
Shooting Three to Make Tutu
Much Ado About Nana
Ex Marks the Spot
To Build or Not to Build
Birth of a Hot Rod
The Great Race
Maybe, Baby
Aisle See You in My Dreams
This Joke's for You
A Sew, Sew Evening
Arrivederci, Binford
Crazy for You
Blow Up
Be True to Your Tool
Dollars and Sense
A Frozen Moment
Feud for Thought
'Twas the Blight Before Christmas
Slip Sleddin' Away
Dream On
Reel Men
The Colonel
Room for Change
The Eve of Construction
Too Many Cooks
It Was the Best of Tims, It Was the Worst of Tims
Fifth Anniversary
Swing Time
What You See is What You Get
Reality Bytes
The Great Race II
Back in the Saddle Shoes Again
Don't Tell Momma
Death Begins at Forty
The Eyes Don't Have It
He Ain't Heavy, He's Just Irresponsible
Borland Ambition
Let's Go to the Videotape
Quibbling Siblings

My Dinner with Wilson

Ye Olde Shoppe Teacher

Some Like It Hot Rod

'Twas the Night Before Chaos

The Route of All Evil

Brother, Can You Spare a Hot Rod?

Superbowl Fever

Bachelor of the Year

It's My Party

A House Divided

The Naked Truth

Talk to Me

No, No, Godot

Tool Time After Dark - Part 1

Tool Time After Dark - Part 2

Sisters and Brothers

A Marked Man

Wilson's Girlfriend

A Taylor Runs Through It

The First Temptation of Tim

Her Cheatin' Mind

Jill's Surprise Party

Advise and Repent

Let Them Eat Cake

The Look

Room Without a View

Chicago Hope

Doctor in the House

That's My Momma

'Twas The Flight Before Christmas

Oh, Brother

High School Confidential

Tanks for the Memories

The Vasectomy One

Fear of Flying

When Harry Kept Delores

Eye on Tim

The Bud Bowl

Engine and a Haircut, Two Fights

The Longest Day

Mr. Wilson's Opus

Shopping Around

Alarmed by Burglars

Games, Flames and Automobiles

At Sea
Future Shock
Workshop 'Til You Drop
Burnin' Love
Al's Video
Whose Car is it Anyway?
I was a Teenage Taylor
Jill and Her Sisters
The Toolman Delivers
The Wood, the Bad and the Hungry
Workin' Man Blues
No Place Like Home
The Flirting Game
The Karate Kid Returns
Totally Tool Time
A Funny Valentine
Wilson's World
Something Old, Someone Blue
Communication Breakdown
My Son, the Driver
Insult to Injury
Family Unties
The Feminine Mistake
Taps
The Kiss & the Kiss-Off
Quest for Fire
The Clash of the Taylors
Room at the Top
Pump You Up
A Night to Dismember
The Niece
Jill's Passion
Losing My Religion
Thanksgiving
The Dating Game
Bright Christmas
Old College Try
An Older Woman
Tim 'The Landlord' Taylor
Say Goodnight, Gracie
What a Drag
Taking Jill for Granite
Futile Attraction
Desperately Seeking Willow

The Write Stuff
The Son Also Mooches
Believe It or Not
Rebel Without Night Driving Privileges
Tool-Thousand-One: A Space Odyssey
From Top to Bottom
Whitewater
Adios
All in the Family
Taylor Got Game
Al's Fair Lady
Bewitched
Not So Great Scott
Tim's First Car
Mr. Likeable
Thanks, But No Thanks
Home for the Holidays
Ploys for Tots
Chop Shop 'Til You Drop
Home Alone
Knee Deep
Mark's Big Break
Young at Heart
Love's Labor Lost, Part I
Love's Labor Lost, Part II
Neighbors
A Hardware Habit to Break
Loose Lips and Freudian Slips
Trouble-a-Bruin
Dead Weight
The Long and Winding Road, Part I
The Long and Winding Road, Part II
The Long and Winding Road, Part III
Mow Better Blues
Off Sides
Satellite on a Hot Tim's Roof
Wild Kingdom
Adventures in Fine Dining
Nothing More Than Feelings
Bubble, Bubble, Toil and Trouble

MODEL TALENT AGENCIES

California

Artist Management 835 Fifth Avenue, Suite 411 San Diego, CA 92101-6137 Tel - 619.233.6655	Bordeuxl 616 N. Robertson Blvd W. Hollywood, CA 90069 Tel - 310.289.2550	Click 9057 Nemo Street N. Hollywood, CA 90096 Tel - 310.246.0080
Extraordinaire Models & Talent 200 New Stine Road Bakersfield, CA 93309 Tel - 661.397.4440	Industry Model & Talent 942 Market Street, Suite 506 San Francisco, CA 94102 Tel - 415.986.6151	Julie Nation Academy Talent 2455 Bennett Valley Rd 110A Santa Rosa, CA 95404 Tel - 707.575.8585
LA Models 8335 Sunset Blvd Los Angeles, CA 90211 Tel - 213.656.9572	Look Models & Talent 166 Geary St Suite 1400 San Francisco, CA 94108 Tel - 415.781.2822	Mari Smith Model & Talent 101 State Place, Suite D Escondido, CA 92029 Tel - 760.745.1627
Colleen Cler Agency 178 S. Victory Blvd, Suite 108 Burbank, CA 91502 Tel - 818.841.7943	Marla Dell Talent 2124 Union Street, Suite C San Francisco, CA 94123 Tel - 415.563.9213	McCright Talent Agency 1011 Stine Road Bakersfield, CA 93309 Tel - 661.835.1305
Midwest Talent Management 4821 Lankershim Blvd suite F 149 N. Hollywood, CA 91601 Tel - 818.765.3785	Panda Talent Agency 3721 Hoen Avenue Santa Rosa, CA 95405 Tel - 707.576.0711	Tina Real Casting 3108 5th Ave, Suite C San Diego, CA 92103 Tel - 619.298.0544
TMA Agency 129 W Wilson Street, Suite 202 Costa Mesa, CA 92627 Tel - 949.574.1100	Ford Models Agency 88826 Burton Way Beverly Hills, CA 90211 Tel - 310.276.8100	Stars The Agency 23 Grant Ave, 4th Fl San Francisco, CA 94108 Tel - 415.421.6272
Cunningham, Escott & Depene 10635 Santa Monica Blvd., Suite 140 Los Angeles, CA 90025 Tel - 310.475.2111	Look Model 166 Geary St 14th Floor San Francisco, CA 94108 Tel - 415.781.5665	Mitchel Model Management 323 Geary St, Suite 303 San Francisco, CA 94102 Tel - 415.395.9291

Alabama

Cathi Larsen Model & Talent Agency 1675 Montclair Road, Suite 136 Birmingham, AL 35210 Tel - 205.951.2445	Dimension Scouting Network 331 West 9th Street Oxford, AL 36203 Tel - 256. 831.0046	Dimension Scouting Network 331 West 9th Street Oxford, AL 36203 Tel - 256. 831.0046
E'lan Agency 1446 Montgomery Highway Birmingham, AL 35216 Tel - 205.823.9180	Pazazz Models 4367 Downtowner Loop N Ste. A Mobile, AL 36609 Tel - 334.343.2502	Rare Quality Models & Talent P.O. Box 1545 Dothan, AL 36302 Tel - 334.671.2200
Ready for the World Modeling 3910 Llyde Ln. Montgomery, AL 36116 Tel - 334.271.6817	Real People Models & Talent Inc. 1827 1st Ave N. Suite 209 Birmingham, AL 35233 Tel - 205.323.5437	Studio 4 Models 1202 Martin Gap Rd Albertsville, AL 35951 Tel - 256.878.0910

Talentscouts Model & Talent 5928 Shane Circle Pinson, AL 35126 Tel - 205.681.5889		

Florida

Adria Phyllise Model & Talent 1939 W. Green St Tampa, FL 33607 Tel - 813.253.6555	Alexa Model & Talent Management 4100 West Kennedy Blvd Tampa, FL 33609 Tel - 813.289.8020	Amazing Kids Inc. 420 Lincoln Road, Suite 330 Miami Beach, FL 33139 Tel - 305.532.1118
Arthur Arthur 245 SE. 1st St. Suite 212 Miami Fl 33131 Tel - 813.645.9700	Avenue Productions Inc. 2810 E Oakland Park Blvd, Ste 308 Ft. Lauderdale, FL 33306 Tel - 954.561.1226	Azuree Talent Agency Inc. 1115 Kentucky Ave Winter Park, FL 32789 Tel - 407.629.5025
Bella Model & Talent Management 8380 Bay Meadows Road, Suite 14 Jacksonville, FL 32256 Tel - 904.739.0619	Boom Model & Talent Agency 13012 N Dale Mabry Tampa, FL 33618 Tel - 813.264.1373	Elite Miami 1200 Collins Avenue, Suite 207 Miami Beach, FL 33139 Tel - 305.674.9500
Famous Faces Entertainment 2013 Harding Street Hollywood, FL 33020 Tel - 954.922.0700	Ford Models 1775 Collins Ave Suite 216 Miami Beach, FL 33139 Tel - 305.534.7200	Irene Marie 728 Ocean Drive, Miami Beach, FL 33139 Tel - 305.672.2929
It's Totally You Inc. 4427 SE 16th Place, Suite 3 Cape Coral, FL 33904 Tel - 239.541.9111	MMG 846 Lincoln Rd 7th Fl. Miami Beach Fl. 33139 Tel - 305.672.8300	Martin & Donald's Talent Agency 2131 Hollywood Blvd., Suite 306 Hollywood, FL 33029 Tel - 954.921.2427
Models Exchange Talent Agency 2425 E Commercial Boulevard Ft. Lauderdale, FL 33308 Tel - 954.491.1014	Next Management 1688 Meridian Avenue, Suite 800 Miami Beach, FL 33139 Tel - 305.531.5100	Portfolio Model Mgmt & Talent 235 Lincoln Road, Suite 201/207 Miami Beach, FL 33139 Tel - 305.534.4210
The Hurt Agency 400 N. New York Ave Suite 207 Winter Park, Fl 32789 Tel - 407.740.5700	Turnabout Talent Agency 333 SE Tressler Drive, Suite G Stuart, FL 34994 Tel - 561.283.1449	Wilhelmina Models • Miami 927 Lincoln Road, Suite 200 Miami Beach, FL 33139 Tel - 305.672.9344
World Of Kids 1460 Ocean Drive, Suite 205 Miami Beach, FL 33139 Tel - 305.672.5437	The Green Agency 1329 Alton Rd Miami Beach, FL 33139 Tel - 305.532.9225	

Georgia

About faces Model & talent 3400 Peachtree Rd Suite 147 Atlanta GA 30326 Tel - 404.233.2006	Ah! Annie Harvey Talent 77 E Andrews Drive NW, Suite 243 Atlanta, GA 30305 Tel - 404.266.0427	Arlene Wilson Model Management 887 W Marietta Street NW #101 Atlanta, GA 30318 Tel - 404.876.8555
Atlanta Models & Talent 2970 Peachtree Road NW, Suite 660 Atlanta, GA 30305 Tel - 404.261.9627	Atlanta's Young Faces/Look Model 6075 Roswell Road NE, Suite 118 Atlanta, GA 30328 Tel - 404.255.3080	Mademoiselles International 2901 University Avenue Columbus, GA 31907 Tel - 706.561.9449

Madison Agency 426 Marietta Street NW, Suite 410 Atlanta, GA 30313 Tel - 678.302.8650	Babes 'N Beaus 4757 Canton Road, Suite 107 Marietta, GA 30066 Tel - 770.928.5832	The Burns Agency 3800 Brettonwoods Rd. Decatur, GA 30032 Tel- 404.303.8995
Dance 411 Productions 5825 Glenridge Dr NE Bldg 2 #106A Atlanta GA 30328 Tel - 404.843.0033	Elite Model Management 1708 Peachtree St Atlanta, GA 30309 Tel - 404.872.7444	Genesis 1465 Northside Dr. Ste. 120 Atlanta, GA 30318 Tel - 404.350.9212
Houghton Talent 919 Collier Rd, NW Atlanta GA 30318 Tel - 404.603.9454	Lorren & Macy's Modeling Agency P.O. Box 3025 Rome, GA 30164-3025 Tel - 706.235.1175	Millie Lewis Models and Talent 7011 Hodson Memorial Dr Savannah, GA 31406 Tel - 912.354.9525
The Peoples Store/ Hot shot Kids 2004 Rockledge Road NE Atlanta, GA 30324 Tel - 404.874-6448	Real People Models & Talents Inc 1479 Spring Street Atlanta, GA 30309 Tel - 404.872.8582	Ted Borden & Associates P.O. Box 11590 Atlanta, GA 30355 Tel - 404.266.0664
Tara Modeling & Talent Agency 696 Mt Zion Rd Suite C2 Atlanta GA, 30236 Tel - 770.968.7700	The William Reynolds Agency 1936 N. Druid Hills Rd Suite 101 Atlanta GA, 30319 Tel - 404.636.1974	TMA Talents 1702 Dunwoody Place Atlanta, GA 30324 Tel - 404.231.1778

Illinois

Ambassador Talent Agents, Inc 333 N. Michigan Avenue, Suite 910 Chicago, IL 60601 Tel - 312.641.3491	Aria Talent / Ford Models 1017 West Washington, Suite 2C Chicago, IL 60607 Tel - 312.243.9400	Arlene Wilson Model Management 430 W Erie Street Chicago, IL 60610 Tel - 312.573.0200
ASAP Models &Talent Ltd P.O. Box 408080 Chicago, IL 60640 Tel - 773.755.0000	Baker & Rowley Talent 1347 W. Washington 1B Chicago, IL 60607 Tel - 312.850.4700	Beebe's Babies 2719 Windy Drive Naperville, IL 60565 Tel - 630.961.2424
Best Faces of Chicago 1152 North La Salle, Suite F Chicago, IL 60610 Tel - 312.944.3009	Best Impressions Model & Talent 477 E Butterfield Rd, Suite 302 Lombard, IL 60148 Tel - 630.434.2214	Claire Model and Talent P.O. Box 1028, Wheeling, IL. 60090 Tel - 847.459.4242
Elite Model Management 58 West. Huron Chicago, IL 60610 Tel - 312.943.3226	Encore Talent 7000 N. Sacramento Blvd Chicago, IL 60612 Tel - 773.638.7300	Ford Models, Inc. 641 W. Lake St., Suite 402 Chicago, IL 60661 Tel - 312.707.8515
Geddes Agency 1633 Halsted Av, Suite 400 Chicago, IL 60614 Tel - 312.787.8333	Lily's 1301 W. Washington, Suite B Chicago, IL 60607 Tel 312.601.2345	McBlaine & Associates 805 Touhy Ave Park Ridge, IL 60068 Tel - 847.823.3877
Models Unlimited LaSalle St., Ste 202 Chicago, IL 60610 Tel - 312.329.1001	Nouvelle Talent P.O. Box 578100 Chicago, IL 60657 Tel - 312.944.1133	Royal Model Management 1051 Perimeter Dr Schaumburg, IL 60173 Tel - 847. 240.4216

Shawnee Studios Model Management 102 W. Main St. Mt Olive, IL 62069 Tel - 217.999.2522	Stewart Talent Management 58 West. Huron Chicago, IL 60610 Tel - 312.943.0892	AS Management 360 Killian Hill Road, Suite F-2 Lilburn, GA 30047 Tel - 770.806.1223
One Magnificent Modes Inc P.M.B. 411 333 W. North Av. Chicago, IL 60610 Tel - 312.475.1447	Legende Talent & Model Mgmt One E. Huron St Chicago, IL 60611 Tel - 312.640.0777	Shirley Hamilton 333 E. Ontario Chicago, IL 60611 Tel - 312.787.4700

Maryland

Dazzling Stars Model Mgmt. Inc 1726 Reisterstown Rd, Suite 210 Pikesville, MD 21208 Tel - 410.486.7766	Flair 5602 Baltimore National Pike Catonsville, MD 21228 Tel - 410.744.3901	King Productions Model & Talent 8900 Edgeworth Drive Suite B Capitol Heights, MD 20743 Tel - 301.350.9778
Nova Models, Inc. 2120 N Charles Street Baltimore, MD 21218 Tel - 410.752.6682	Ramp Model Management 1713 Park Avenue, Suite B Baltimore, MD 21217 Tel - 410.737.0000	Media Star Promotions 6305 Blair Hill Lane Baltimore, MD 21209 Tel - 877.710.2847
MTM Management 7801 York Rd, Suite 305 Towson, MD 21204 Tel 410.821.6661		

Massachusetts

The Models group 374 Congress St . #305 Boston MA, 02210 Tel - 617.426.4711		

Minnesota

Caryn Model & Talent Agency 100 N 6th Street, Suite 270B Minneapolis, MN 55403 Tel - 612.349.3600	Meredith Model & Talent Agency 800 Washington Ave. North # 511 Minneapolis, MN 55101 Tel - 612.340.9555	Model Ink Avenue 1193 Earl Street St Paul, MN 55106 Tel - 651.772.1670
Moore Creative Talent Inc 1610 W Lake Street Minneapolis, MN 55408 Tel - 612.827.3823	MTM Agency 1021 Bandana Boulevard, Ste 128 St Paul, MN 55108 Tel - 651.642.1222	Wehmann Models & Talent # 202 1128 Harmon Place Minneapolis, MN 55403 Tel - 612.333.6393
Moore Creative Talent 1610 W. Lake Street Minneapolis, MN 55408 Tel. - 612.827.3823	Marey Studio, Inc. 3450 Bonna Delle Ct St. Cloud, MN 56301 Tel - 320.259.1304	Kids.com 5037 France Ave South Minneapolis, MN 55410 Tel - 952.285.2901

New Jersey

AC Talent/McCullough Models 8 S Hanover Avenue Atlantic City, NJ 08402 Tel - 609.822.2222	Axis Model & Talent 46 Church Street Montclair, NJ 07042 Tel - 973.783.4900	Cover Girl Studio Model Mgmt P.O. Box 222 River Edge, NJ 07661 Tel - 201.261.2042
Jo Anderson Models 1 Eves Drive, Suite 111 Marlton, NJ 08053 Tel - 856.596.7200	Kathy Donahue's Models/Actors 383 N Kings Highway, Suite 5B Cherry Hill, NJ 08034 Tel - 856.482.9113	Meredith Model & Talent Agency 10 Furler Street Totowa, NJ 07512 Tel - 973.812.0122
Model Team 55 Central Avenue Ocean Grove, NJ 07756 Tel - 732.988.3648	Models on the Move 1200 Rte 70, Barclay Towers, Suite6 Cherry Hill, NJ 08034 Tel - 856.667.1060	Kids.com 343 Passaic Ave Fairfield, NJ 07004 Tel - 973.575.7300

New York

American 155 Spring St New York, New York 10012 Tel - 212.941.5858	Abrams Artist & Associates 420 Madison Ave New York, NY 10017 Tel - 212.953.8980	Boss Models One Ganesvoort St New York, NY 10014 Tel - 212. 242.2444
Carson-Adler Agency 250 W. 57th St New York, NY 10107 Tel - 212.307.1882	Cunningham, Escott & Dipene 257 Park Avenue South, Suite 900 New York, NY 10010 Tel - 212.477.1666	Elite Model Management 111 East 22nd St., 2nd Fl New York, NY 10010 Tel - 212.529.9700
Fifi Oscard Agency 24 West 40th St., 17th Fl. New York, NY 10018 Tel - 212.764.1100	Ford Models 142 Greene St, 4th Fl. New York, NY 10012 Tel - 212.966.3565	Grace Del Marco 350 Fifth Ave, Ste., 3110 New York, NY 10118 Tel - 212.629.6404
Ikon 140 West 22nd St. 2nd Fl New York, NY 10011 Tel - 212 691.2363	Karin Models 6 West 14th St, 3rd Fl New York, NY 10011 Tel - 212.226.4100	Next Management 15 Watts Street, 5th Fl New York, NY 10013 Tel - 212.925.5100
Nouvelle Talent 20 Bethune St Suite 4A New York, New York 10014 Tel - 212.645.0940	The Lyons Group 505 Eight Ave, Fl 12A New York, NY 10018 Tel - 212.239.3539	Wilhelmina International 300 Park Ave, South, 2nd Fl New York, NY 10010 Tel - 212.473.1253\
Wilhelmina 300 Park Ave S. New York, New York 10012 Tel - 212.473.0700	William Morris Agency 1325 Avenue of the Americas New York, NY 10019 Tel - 212.586.5100	Micheal Amato 1650 Broadway St. Ste 307 New York, NY 10019 Tel - 212.247.4456
Zoli Model Management 3 West 18th St, 5th Fl. New York, NY 10011 Tel - 212.242.1500	Dorothy Palmer Agency 235 West 56th St., Suite 24K New York, NY 10019 Tel - 212.765.4280	A Plus 18 West 21st Street, 7th Fl New York, NY 10010 Tel. - 212.633.8848
Diva & Adonis 451 Greenwich St, 5th Fl New York, NY 10013 Tel - 212.807.0055	Thompson Model & Talent 50 West 34th St., Ste. 6C6 New York, NY 10001 Tel - 212.947.6711	Kids Power Model & Talent 109 W. 38th St New York, NY 10018 Tel - 212.768.0970

North Carolina

Actors & Entertainers P.O. Box 486 Kernersville, NC 27285 Tel - 336.993.1611	Artist Resources Agency, Inc 174 Hunters Glen Dr Summerfield NC 27358 Tel - 336.349.6167	The Brock Agency 329 13th Avenue NW Hickory, NC 28638 Tel - 828.322.8553
The Burns Agency 1255 Marlborough Ln. Winston-Salem, NC 27105 Tel- 336.744.5037	Capital Artists Of North Carolina 1405 BloomingdaleDr Cary, NC 27511 Tel - 919.467.8682	Carolina Talent P.O. Box 36422 Charlotte, NC 28236 Tel - 704.332.3218
Directions USA 3717C West Market Street Greensboro, NC 27403 Tel - 336.292.2800	Evolution Models & Talent 310 Arlington Ave Suite 403 Charlotte NC 28203 Tel - 704.375.9797	ICE Agency 8318 Pineville Matthews Rd Ste. 265 Charlotte, NC 28226 Tel - 704.543.4120
JTA Talent Inc 820 East Boulevard Charlotte, NC 28203 Tel - 704.377.5987	Kids Unlimited Talent 1816 Bonnibee Ct Raleigh, NC 27612 Tel - 919.785.1223	Libby Stone Modeling School 1819 Charlotte Drive Charlotte, NC 28203 Tel - 704.377.9299
Marilyn's Model & Talent Agency 601 Norwalk Street Greensboro, NC 27407 Tel - 336.292.5950	Maultsby Talent P.O. Box 11039 Wilmington, NC 28404 Tel - 910.509.4008	M3 Model Management P.O. Box 3003 Charlotte, NC 28230 Tel - 704.492.5235
Quick Talent & Assocaites 305 South 7th Street Wilmington, NC 28401 919 762.9273	SASS Modeling & Talent Agency 8419 Clear Meadow Ln Charlotte, NC 28227 Tel - 704.545.5411	Talent Connection 338 N Elm Street, Suite 204 Greensboro, NC 27401 Tel - 336.274.2499
Talent One 7125 cheviot Hills Dr. Raleigh, NC 27616 Tel - 919.8724828	Talent Trek 825-C Merrimon Ave, PMB 356 Ashville, NC 28804 Tel - 828.251.0173	Talent Link Talent & Modeling P.O. Box 77185 Charlotte, NC 28271 Tel - 704.335.0027
Tommi Long Jones Talent Agency 320 Settlers Lane Charlotte, NC 28202 Tel - 704.335.0388	William Pettit Enterprises P.O. Box 11798 Charlotte, NC 28220 Tel-704.643.8880	

Ohio

Ashley Talent Agency 10948 Reading Road, Suite 310 Cincinnati, OH 45241 Tel - 513.554.4836	Cam Talent 1150 W 8th Street, Suite 262 Cincinnati, OH 45203 Tel - 513.421.1795	Cam Talent 1350 W 5th Avenue, Suite 25 Columbus, OH 43212 Tel - 614.488.1122
Cincinnati Model Agency 6047 Montgomery Road Cincinnati, OH 45213 Tel - 513.351.2700	D'Avila Model & Talent Management 5840 Ridge Road Cleveland, OH 44129 Tel - 440.843.7200	The Heyman Talent Agency 3308 Brotherton Rd Cincinnati, OH 45209 Tel - 513.533.3113
Docherty Agency 2044 Euclid Ave Suite 500 Cleveland, OH 44115 Tel - 216.522.1300	The Heyman Talent Agency 1612 S. High St Columbus, OH 43207 Tel - 614.542.0882	IMI Talent Management 9700 Rockside Road, Suite 410 Cleveland, OH 44125 Tel - 216.901.9710

Roof - Goenner Talent Agency 123 Webster St 2nd level Dayton, OH 45402 Tel - 937.586.0799	Pro Model Management 3296 W. Market Street Akron, OH 44333 Tel - 330.867.4125	S2 Management Group 545 W. Nationwide Blvd. Columbus, OH 43215 Tel - 614.294.0100
The Right Direction Model Mgmt. 4770 Indianola Ave, Suite 160 Columbus, OH 43214 Tel - 614.848.3357	Wings Model Management 906 Main St, Suite 207 Cincinnati, OH 45202 Tel - 513.929.9464	

Pennsylvania

Bowman Agency P.O. Box 4071 Lancaster, PA 17604 Tel - 717.898.7716	Click Models 216 Green Tree Street West Chester, PA 19382 Tel - 610.399.0700	Donatelli Model & Casting Agency 156 Madison Avenue Reading, PA 19605 Tel - 610.921.0777
Expressions Agency 220 Church Street Philadelphia, PA 19106 Tel - 215.923.4420	Fashion Mystique Modeling Agency 611 N Mountain Road Harrisburg, PA 17112 Tel - 717.561.2099	Docherty Agency 109 Market St, 3rd Fl Pittsburg, PA 15222 Tel - 412.765.1400
Main Line Models 1215 W Baltimore Pike, Suite 9 Media, PA 19063 Tel - 610.565.5445	Marilyn E Kane Model Mgmt 1022 N Main Street Butler, PA 16001 Tel - 724.287.0576	Plaza 7 Model & Talent Reps 160 North Gulph Road King of Prussia, PA 19406 Tel - 610.337.2693
Prestige Modeling 10028 Frankstown Road Pittsburgh, PA 15235 Tel - 412.826.0740	Pro-Model Management 1244 Hamilton Street Allentown, PA 18102 Tel - 610.820.5359	Reinhard Model & Talent Agency 2021 Arch Street Philadelphia, PA 19103 Tel - 215.567.2008
Slickis Models 1777 Walton Road, Suite 204 Blue Bell, PA 19422 Tel - 215.540.0440	The Talent Group, Inc. 2820 Smallman Street Pittsburgh, PA 15222 Tel - 412.471.8011	

South Carolina

Betty Lane Models Talent Agency 951 Doyle Street Orangeburg, SC 29115 Tel - 803.534.9672	Z-1 Models & Talent 141 Gadsden Street Chester, SC 29706 Tel - 803.581.2278	Carmicheal Model & Talent Po Box 884 Johnsonville, SC 29555 Tel- 843.386.3320
Millie Lewis of Columbia 3612 Landmark Drive, Suite D Columbia, SC 29204 Tel - 803.782.7338	Millie Lewis Models & Talent 7475 Northside Dr. N. Charleston, SC 29402 Tel - 843.571.7781	Millie Lewis Models & Talent 1228 S. Pleasentburg Dr. Greenville, SC 29605 Tel - 864.299.1101
Shaws Model & Talent 200 Berkshire Dr Columbia, SC 29223 Tel- 803.699.0158	Showcase Models & Talent 1200 33rd Avenue South N Myrtle Beach, SC 29582 Tel - 843.272.8009	

Tennessee

Advantage Models & Talent 230 Franklin Rd. Suite 802 Franklin, TN 37064 Tel - 615.790.5001	AMAX Talent Agency 2815 Bransford Ave Nashville, TN 37204 Tel - 615.292.0246	Ambiance Models & Talent 1096 Dayton Boulevard Chattanooga, TN 37405 Tel - 423.265.2121
Billy Deaton Talent 1300 Division Street, Suite 102 Nashville, TN 37203 Tel - 615.244.4259	Capitol Management & Talent Group 1300 Division Street, Suite 200 Nashville, TN 37203 Tel - 615.244.2440	Coleman Model & Talent Agency 417 E. Iris Dr. Nashville, TN 37204 Tel - 615.385.5797
Colors Agency, Inc 1985 Madison Ave Suite 3 Memphis, TN 38104 Tel - 901.726.9300	Heartbeat Productions 832 S. Cooper St. Memphis TN 38104 Tel - 901.278.0138	Jo-Susan Model & Talent 2817 W End Avenue, Suite 210 Nashville, TN 37203 Tel - 615.327.8726
Moore/Higgs Casting 417 Welshwood Drive Nashville, TN 37211 Tel - 615.831.0039	Richmond Casting Group 1719 Westend Avenue Nashville, TN 37203 Tel - 615.320.9594	Robbins Model & Talent 176 Walnut Bend Memphis, TN 38018 Tel - 901.753.8360
18 Karat Talent & Modeling Agency 6409 Deane Hill Dr Knoxville TN 37919 Tel - 865.558.0004	The Talent Trek Agency 5401 Kingston Pike Knoxville TN 37919 Tel - 865.977.8735	The Donna Groff Agency Inc P.O. Box 382517 Memphis, TN 38183-2517 Tel - 901.854.5561
The Hurd Agency 500 Eversholt Court Antioch, TN 37013 Tel - 615.399.9901	TML Talent Agency P.O. Box 40763 Nashville, TN 37204 Tel - 615.321.5596	Zoom Agency 1890 N. Germantown Parkway #104 Cordova TN 38016 Tel - 901.842.6682

Texas

Avant Models & Casting, Inc. 6601 Blanco Rd Suite 200 San Antonio, TX 78216 Tel - 210.308.8411	Calliope Talent & Modeling Agency 1802 NE Loop 410, Suite 107 San Antonio, TX 78217 Tel - 804.1055	Campbell Agency 3906 Lemmon Ave Suite 200 Dallas TX 75219 Tel - 214.522.8991
Condra /Artista Model & Talent 10737 Gulfdale Rd Suite 200 San Antonio, TX 78216 Tel - 210.492.9947	DB Talent 7402 Brodie Ln. Austin, TX 78745 Tel - 512.292.1030	Diane Dick Int. Modeling & Talent 1410 S Washington Street Amarillo, TX 79102 Tel - 806.376.8736
First Models & Talent Agency 5433 Westheimer Avenue, Suite 310 Houston, TX 77056 Tel - 713.850.9611	Kim Dawson Agency Inc-Talent Div. 2710 N Stemmons Freeway, Ste 700 Dallas, TX 75207 Tel - 214.630.5161	Mary Collins Agency 2909 Cole Avenue, Suite 250 Dallas, TX 75204 Tel - 214.871.8900
Page Parkes Agency 3303 Lee Parkway Suite 205 Dallas TX 75258 Tel - 214.526.4434	Pastorini-Bosby Talent Agency 3013 Fountainview Drive, Suite 240 Houston, TX 77057-6120 Tel - 713.266.4488	Reflections Fashion & Talent Agency 110 Alta Plaza Corpus Christi, TX 78411 Tel - 361.854.9277

Sherry Young/Mad Hatter 2620 Fountainview, Suite 212 Houston, TX 77057 Tel - 713.266.5800	Talents International 9449 Briar Forest, Suite 3105 Houston, TX 77063 Tel - 713.780.4477	The Talent House 812 North Virginia Street, Suite 204 El Paso, TX 79902 Tel - 915.533.1945
The William Ware Agency 2508 Valleyview Ln Dallas, TX 75234 Tel - 214.638.6200	Tomas Agency 14275 Midway Road, Suite 220 Addison, TX 75001 Tel - 972.687.9181	Neal Hamil Agency 7887 San Felipe Ste. 227 Houston, TX 77063 Tel - 713.789.1335
The Campbell Agency 3906 Lemmon Ave, Suite 200 Dallas, TX 75219 Tel - 214.522.8991		

Virginia

Beaute 2816 Taxton Ln. Oakton, VA 22124 Tel -703.734.1131	Beautiful People Model and Talent 3237 Tyreneck Rd. Portsmouth, VA 23703 Tel - 757.686.5626	Bonnie Adleman Agency Inc. Midlothian VA 23113 2600 Robious Crossing Dr Tel - 804.379.8200
Encore! 17 Keith's Lane Alexandria, VA 22314 Tel - 703.548.0900	Evie Mansfield Modeling Agency 505 S. Independence Blvd., Ste 205 Virginia Beach, VA 23452 Tel - 757.490.5990	Glamour Modeling & Talent 1115 Independence Blvd. Suite 110 Virginia Beach, VA 23455 Tel - 757.363.8844
Marilyn's Model & Talent 2721 Charstone Dr. Midlothian, VA 23113 Tel - 804.379.1946	Modelogic Inc. 2501 E Broad St Richmond VA 23223 Tel - 804.644.1000	Millennium Model & Talent Mgmt. 8321 Old Courthouse Rd Vienna, VA 22182 Tel - 703.903.0040
Model and Talent Management 249 South Van Dorn St Alexandria, VA 22304 Tel - 703.379.2555	New Faces Models 8230 Leesburg Pike, Suite 520 Vienna, VA 22182 Tel - 703.821.0786	On Call Models & Talent 946 Ferryman Quay Chesapeake, VA 23323 Tel - 757.485.1201
Studio M 1709 Wainwright Dr Reston, VA 20190 Tel - 703.471.4666	Talent Link P.O Box 12007 Newport News VA 23612 Tel - 757.249.2232	Uptown Talent P.O Box 29388 Richmond VA 23432 Tel - 804.740.0307

Washington, DC

Anne Schwab's Model Store 906 D Street NE Washington, DC 20002 Tel - 202.333.3560	Doran Model & Talent Agency 1301 Massachusetts Av NW, Ste 701 Washington, DC 20005 Tel - 202.333.6367	

KIDS MODEL TALENT AGENCIES

ALABAMA TALENT & QUALITY MODELS
PO BOX 020198
TUSCALOOSA, AL 35402

ALABAMA TALENT MANAGEMENT
PO BOX 020198
TUSCALOOSA, AL 35402-0198

BAREFOOT MODELS & TALENT
750 DOWNTOWN LOOP W., STE. G
MOBILE, AL 36609

CATHI LARSEN MODEL & TALENT AGENCY
1675 MONT CLAIRE RD., STE. 136
BIRMINGHAM, AL 35210

CYNTHIA'S MODEL & TALENT
2030 4TH ST. EAST
MOTGOMERY, AL 36106

E'LAN AGENCY
1446 MONTGOMERY HIGHWAY
BIRMINGHAM, AL 35216

MACY'S MODELING SCHOOL & AGENCY
994 MAIN STREET
ROANOKE, AL 36276

PAMA AGENCY
721 CLINTON AVENUE, W. STE 4
HUNTSVILLE, AL 35801

PAZAZZ MODELS
4367 DOWNTOWNER LOOP N STE A
NOBILE, AL 36609

PRATER & PRATER
2642 O'NEAL CIRCLE
BUIRMINGHAM, AL 35226

READY FOR THE WORLD MODELING
4429 TROY HIGHWAY
MONTGOMRY, AL 36116

REAL PEOPLE MODELS & TALENT INC.
714 32ND STREET S
BIRMINGHAM, AL 35233

STUDIO 4 MODELS
1202 MARTIN GAP RD
ALBERTSVILLE, AL 35951

TALENTSCOUTS MODEL & TALENT
5928 SHANE CIRCLE
PINSON, AL 35951

ARTIST MANAGEMENT
835 FIFTH AVENUE, STE. 411
SAN DIEGO, CA 92101-6137

BELLISSIMA MODEL & TALENT MGT.
1055 W. COLLEGE AVE.
SANTA ROSA, CA 95401

BOBBY BALL
4342 LANKERSHIM BLVD.
LOS ANGELES, CA 91602

ELEGANCE TALENT & MODEL MGT.
2763 STATE STREET
CARLSBAD, CA 92008

EXTRAORDINAIRE MODELS & TALENT
200 NEW STINE ROAD
BAKERSFIELD, CA 93309

INDUSTRY MODEL & TALENT
942 MARKET STREET, STE. 506
SAN FRANCISCO, CA 94102

JULIE NATION ACADEMY TALENT
2455 BENNETT VALLEY RD 110A
SANTAW ROSA, CA 95404

MARI SMITH MODEL & TALENT
101 STATE PLACE, STE. D
ESCONDIDO, CA 92029

MARLA DELL TALENT
2124 UNION STREET, STE. C
SAN FRANCISCO, CA 94123

MAX MODEL & TALENT MANAGEMENT
980 SPRINGFIELD STREET
COSTA MESA, CA 92626

McCRIGHT TALENT AGENCY
1011 STINE ROAD
BAKERSFIELD, CA 93309

PANDA TALENT AGENCY
3721 HOEN AVENUE
SANTA ROSA, CA 95405

THE TALENT CONNECTORS
11211 COHASSET STREET
SUN VALLEY, CA 91352

TINA REAL CASTING
3108 5TH AVE., STE. C
SAN DIEGO, CA 92103

TMA AGENCY
129 W. WILSON ST., STE 202
COSTA MESA, CA 92627

WOW MODEL & TALENT MANAGEMENT
1610 OAKPARK BLVD., STE. 2
PLEASANT HILL, CA 94523

FORD MODELS AGENCY
88826 BURSTON WAY
BEVERLY HILLS, CA 90211

STARS THE AGENCY
23 GRANT AVE., 4TH FL
SAN FRANCISCO, CA 94108

CUNNINGHAM, ESCOTT & DEPENE
10635 SANTA MONICA BLCD., STE. 140
LOS ANGELES, CA 90025

LOOK MODEL
166 GEARY ST. 14TH FL
SAN FRANCISCO, CA 94108

MITCHELL MODEL MANAGEMENT
323 GEARY ST., STE. 303
SAN FRANCISCO, CA 94102

COLLEEN CLER AGENCY
178 S. VICTORY BLVD., STE. 108
BURBANK, CA 91502

ALEXA MODEL & TALENT MGT
4100 WEST KENNEDY
TMPA, FL 33609

AMAZING KIDS INC.
420 LINCOLN RD., STE. 330
MIAMI BEACH, FL 33139

AVENUE PRODUCTIONS INC.
2810 E. OAKLAND PARK BLVD, STE.308
FT. LAUDERDALE, FL 33306

AZUREE TALENT AGENCY INC.
140 N. ORLANDO AVE., STE. 120
WINTER PARK, FL 32789

BELLA MODEL & TALENT MGT.
8380 BAY MEADOWS RD., STE. 14
JACKSONVILLE, FL 32256

BOOM MODELS & TALENT AGENCY
13012 N.DALE MABRY
TAMPA, FL 33618

ELITE MIAMI
1200 COLLINS AVE., STE. 207
MIAMI BEACH, FL 33139

FAMOUS FACES ENTERTAINMENT
2013 HARDING STREET
HOLLYWOOD, FL 33020

FORD MODELS
311 LINCOLN SUITE 205
MIAMI BEACH, FL 33139

IT'S TOTALLY YOU INC.
4427 SE 16TH PLACE, STE. 3
CAPE CORAL, FL 33904

MARTIN & DONALD'S TALENT AGENCY
2131 HOLLYWOOD BLVD., STE. 306
HOLLYWOOD, FL 33029

MODELS EXCHANGE TALENT AGENCY
2425 E COMMERCIAL BLVD.
FT. LAUDERDALE, FL 33308

NEXT MANAGEMENT
1688 MERIDIAN AVE., STE. 800
MIAMA BEACH, FL 33139

PAGE PARKES MODELS REP TALENT
763 COLLINS AVE.
MIAMI BEACH, FL 33139

PORTFOLIO MODEL MGMT & TALENT
235 LINCOLN RD., STE. 201/207
MIAMI BEACH, FL 33139

Sarah Parker Model & Talent
410 Datura Street
W Palm Beach, FL 33401

The Model & Talent Group Inc
102 Park Street
Safety Harbor, Fl 34695

Turnabout Talent Agency
333 SE Tressler Dr., Suite G
Stuart, FL 34994

Wilhelmina Models *Miami
927 Lincoln Rd., Ste. 200
Miami Beach, FL 33139

World Of Kids
1460 Ocean Dr., Ste. 205
Miami Beach, FL 33139

International Artist Group
420 Lincoln Rd., Ste. 382
Miami Beach, FL 33139

The Green Agency
1320 Alton Rd.
Miami Beach, FL 33139

Ah! Annie Harvey Talent
77 E Andrews Dr. NW, Ste. 243
Atlanta, GA 30305

Arlene Wilson Model Management
887 W Marietta St. NW
Atlanta, GA 30318

Atlanta Models & Talent
2970 Peachtree Rd. NW, Ste. 660
Atlanta, GA 30305

Atlanta's Young Faces/Look Model
6075 Roswell Rd. NW, Ste. 118
Atlanta, GA 30328

Atlanta's Winning Edge
360 Killian Hill Rd., Ste. F-2
Lilburn, GA 30047

Babes 'N Beaus
4757 Canton Rd., Ste. 107
Marietta, GA 30066

Axis Model Management
120 W Wieuca Road
Atlanta, GA 30342

Lorren & Macy's Modeling Agency
PO Box 3025
Rome, GA 30164-3025

Mademoiselles International
2901 University Avenue
Columbus, GA 31907

Madison Agency
426 Marietta St., NW, Ste. 410
Atlanta, GA 30313

The Peoples Store
2004 Rockledge Rd. NE
Atlanta, GA 30324

Real People Models & Talents Inc
1479 Spring St.
Atlanta, GA 30309

Slamm Model Management
233 Mitchell St., Ste. 300
Atlanta, GA 30303

Ted Borden & Associates
PO Box 11590
Atlanta, GA 30355

The Talent Group & Hot Shot Kids
3300 Buckeye Rd., Ste. 405
Atlanta, GA 30341

Millie Lewis Models and Talent
7011 Hodson Memorial Dr.
Atlanta, GA 31406

Elite Model Management
1708 Peachtree St.
Atlanta, GA 30309

Genesis
1465 Northside Dr., Ste. 120
Atlanta, GA 30318

Chadz Model and Talent
426 Marietta St. NW, Ste. 410
Atlanta, GA

TMA Talents
1702 Dunwoody Place
Atlanta, GA 30324

Glenn Kennedy Inc. Models & Talent
975 Hunterhill Dr.
Roswell, GA 30075

Ambassador Talent Agents, Inc.
333 N. Michigan Ave. Ste. 910
Chicago, IL 60601

Aria Model & Talent Management
1017 West Washington, Ste. 2C
Chicago, IL 60607

Arlene Wilson Model Management
430 W. Erie St.
Chicago, IL 60610

ASAP Models & Talent LTD
PO Box 408080
Chicago, IL 60640

Beebe's Babies
2719 Windy Dr.
Naperville, IL 60565

Best Faces of Chicago
1152 N. LaSalle, Ste F
Chicago, IL 60610

Claire Model and Talent
PO Box 1028
Wheeling, IL 60090

Classic Model and Talent
225 W. Washington, Ste. 220
Chicago, IL 60606

Emilia Lorence
325 West Huron St., Ste 410
Chicago, IL 60610

Elite Model Management
58 West Huron
Chicago, IL 60610

Models Unlimited
415 N. LaSalle St., Ste. 202
Chicago, IL 60610

Stewart Talent Management
58 West Huron
Chicago, IL 60610

Premiere Model
27 East Monroe St., Ste. 200
Chicago, IL 60603

Royal Model Management
1051 Perimeter Dr.
Schaumburg, IL 60173

Nouvelle Talent
PO Box 578100
Chicago, IL 60657

Shawnee Studios Model Management
102 W. Main St.
Mt Olive, IL 62069

Lily's
1301 W. Washington, Ste. B
Chicago, IL 60607

Elite Model & Talent
58 W. Huron St.
Chicago, IL 60610

Encore Talent
1532 N. Milwaukee Ave., Ste. 204
Chicago, IL 60622

Baker & Rowley Talent
1347 W. Washington 1B
Chicago, IL 60607

Best Impressions Model & Talent
477 E. Butterfield Rd., Ste 302
Lombard, IL 60148

Ford Models, Inc.
641 W. Lake St., Ste. 402
Chicago, IL 60661

McBlaine & Associates
805 Touhy Ave.
Park Ridge, IL 60068

Geddes Agency
1633 Halsted Ave., Ste. 400
Chicago, IL 60614

One Magnificent Modes Inc.
P.M.B. 411
333 W. North Ave.
Chicago, IL 60610

Legende Talent & Model Mgmt
One E. Huron St.
Chicago, IL 60611

Shirley Hamilton
333 E. Ontario
Chicago, IL 60611

ABA Convention Models
4518 Magazine St.
New Orleans, LA 70115

Glamour Modeling & Talent
PO Box 1526
Meraux, LA 70075

Images Model & Talent
200 Polk St., Ste. 101
Lafayette, LA 70501

New Orleans Model/Talent Agency
1347 Magazine St.
New Orleans, LA 70130

Stage 2000 Model & Talent Agency
8133 Royalwood Dr.
Baton Rouge, LA 70806

Victor's International Models
618 N. Carrollton Ave.
New Orleans, LA 70119

Del Corral Model & Talent Agency
130 Telemachus Rd.
New Orleans, LA 70119

The Michael Turner Agency
1805 Line Ave.
Shreveport, LA 71104

Fameagency.com
4518 Magazine St.
New Orleans, LA 70115

Elite Talent & Casting
PO Box 13567
New Orleans, LA 70185

Carpe Diem Model & Talent Agency
2007 Clearview Parkway, Ste. C
Metairrie, LA 70001

About Faces Model & Talent
929 Julia St., 2nd Fl
New Orleans, LA 70113

Dazzling Stars Model Mgmt., Inc.
1726 Reisterstown Rd., Ste 210
Pikesville, MD 21208

Flair
5602 Baltimore National Pike
Catonsville, MD 21228

King Productions Model & Talent
8900 Edgeworth Dr. Ste. B
Capitol Heights, MD 20743

Nova Models Inc.
2120 N. Charles St.
Baltimore, MD 21218

Ramp Model Management
1713 Park Ave., Ste. B
Baltimore, MD 21217

Media Star Promotions
6305 Blair Hill Lane
Baltimore, MD 21209

MTM Management
7801 York Rd., Ste. 305
Towson, MD 21204

Academy for Film & Television
6651 Highway #7
Minneapolis, MN 55426

Caryn Model & Talent Agency
100 N. 6th St., Ste. 2708
Minneapolis, MN 55403

Joe Katz Models
701 4th Ave. S., Ste 1700
Minneapolis, MN 55415

Kaye & Associates Model & Talent
11 4th St., SE, Zumbro
Rochester, MN 55904

Meredith Model & Talent Agency
800 Washington Ave. N
Minneapolis, MN 55101

Model Ink Avenue
1193 Earl St
St. Paul, MN 55106

Moore Creative Talent Inc
1610 W. Lake St.
Minneapolis, MN 55408

MTM Agency
1021 Bandana Blvd., Ste 128
St. Paul, MN 55108

Image I Professional Modeling
118 2nd Ave. North
South Rapids, MN 56379

Welmann Models & Talent
1128 Harmon Place
Minneapolis, MN 55403

Marey Studio, Inc.
3450 Bonna Delle Ct.
St. Cloud, MN 56301

Kids.com
5037 France Ave. South
Minneapolis, MN 55410

AC Talent/McCullough Models
8 S. Hanover Ave.
Atlantic City, NJ 08402

Axis Model & Talent
46 Church St.
Montclair, NJ 07042

Classic Model & Talent Management
87 S. Finley Ave.
Basking Ridge, NJ 07920

Cleri Model Management
402 Main St., Ste. 300
Metuchen, NJ 08840

Cover Girl Studio Model Mgmt
PO Box 222
River Edge, NJ 07661

Jo Anderson Models
1 Eves Dr, Ste. 111
Marlton, NJ 08053

Kathy Donahue's Models/Actors
383 N. Kings Highway, Ste. 5B
Cherry Hill, NJ 08034

Meredith Model & Talent Agency
10 Furler St.
Totowa, NJ 07512

Model Team
55 Central Ave.
Ocean Grove, NJ 07756

Models on the Move
1200 Rte 70, Barclay Towers, Ste. 6
Cherry Hill, NJ 08034

Kids.com
186 Fairfield Rd.
Fairfield, NJ 07004

Nationwide Talent Inc
PO Box 2423
Clifton, NJ 07015

Veronica Goodman Agency
605 W. Rte 70, Ste. 1
Cherry Hill, NJ 08002

Boss Models
One Ganesvoort St.
New York, NY 10014

Grace Del Marco
350 Fifth Ave., Ste. 3110
New York, NY 10118

Elite Model Management
111 East 22nd St., 2nd Fl
New York, NY 10010

Cunningham, Escort & Dipene
257 Park Ave. South, Ste. 900
New York, NY 10010

Ford Models
142 Greene St., 4th Fl.
New York, NY 10022

Ikon
140 West 22nd St., 2nd Fl.
New York, NY 10011

Karin Models
6 West 14th St., 3rd Fl.
New York, NY 10011

The Lyons Group
505 Eight Ave., Fl 12A
New York, NY 10018

McDonalds/Richards
156 Fifth Ave., Ste. 222
New York, NY 10010

Next Management
23 Watts St., 5th Fl.
New York, NY 10013

Carson-Adler Agency
250 W. 57th St.
New York, NY 10107

Wilhelmina International
300 Park Ave., South, 2nd Fl
New York, NY 10010

Zoli Model Management
3 West 18th St., 5th Fl.
New York, NY 10011

Fifi Oscard Agency
24 West 40th St., 17th Fl.
New York, NY 10018

Dorothy Palmer Agency
235 West 56th St., Ste. 24K
New York, NY 10019

Thompson Model & Talent
50 West 34th St., Ste. 6C6
New York, NY 10001

Abrams Artist & Associates
420 Madison Ave.
New York, NY 10017

Kids Power Model & Talent
109 W. 38th St.
New York, NY 10018

William Morris Agency
1325 Avenue of the Americas
New York, NY 10019

Michael Amato
1650 Broadway St., Ste. 307
New York, NY 10019

A Plus
18 West 21st St., 7th FL
New York, NY 10010

Diva & Adonis
451 Greenwich St., 5th Fl.
New York, NY 10013

Artist Resources Agency, Inc
174 Hunters Glen Dr.
Summerfield, NC 27358

The Brock Agency
329 13th Ave. NW
Hickory, NC 28638

Capital Artists Of North Carolina
1405 Bloomingdale Dr.
Cary, NC 27511

Carolina Acting & Talent Scouts
1201 S. Graham St.
Charlotte, NC 28203

Carolina Talent
312 Rensselaer Ave., Ste. 4
Charlotte, NC 28203

CB Talent Management
1253 Colony Dr.
New Bern, NC 28562

Delia Model & Talent Management
1519 N 23rd St., Ste. 203

Directions USA
3717C West Market St.
Greensboro, NC 27403

ICE Agency
8318 Pineville Matthews Rd., Ste. 265
Charlotte, NC 28226

JTA Talent Inc.
820 East Blvd.
Charlotte, NC 28203

Kids Unlimited
1816 Bonnibee Ct.
Raleigh, NC 27612

Libby Stone Modeling School
1819 Charlotte Dr.
Charlotte, NC 28203

Marilyn's Model & Talent Agency
601 Norwalk St.
Greensboro, NC 27407

Maultsby Talent
112 N. Cardinal Dr., Ste. 106
Wilmington, NC 28405

MTM Associates
2650 Park Dr.
Charlotte, NC 28205

On Track Modeling
5500 Executive Center Dr., Ste. 126
Charlotte, NC 28212

SASS Modeling & Talent Agency
5501 Executive Center Dr., Ste, 232
Charlotte, NC 28212

Talent Connection
338 N. Elm St., Ste. 204
Greensboro, NC 27401

Talent Link Talent & Modeling
PO Box 560337
Charlotte, NC 28256

Tommi Long Jones Talent Agency
320 Settlers Lane
Charlotte, NC 28202

Ashley Talent Agency
10948 Reading Rd., Ste. 310
Cincinnati, OH 45241

Cam Talent
1150 W. 8th St., Ste. 262
Cincinnati, OH 45203

Cam Talent
1350 W. 5th Ave., Ste. 25
Columbus, OH 43212

Cincinnati Model Agency
6047 Montgomery Rd.
Cincinnati, OH 45213

D'Avila Model & Talent Management
5840 Ridge Rd.
Cleveland, OH 44129

IMI Talent Management
9700 Rockside Rd., Ste 410
Cleveland, OH 44125

Jo Goenner Talent
2299 State Route 725
Centerville, OH 45455

Taxi Model Management
1300 W. 78th St.
Cleveland, OH 44102

Pro Model Management
3296 W. Market St.
Akron, OH 44333

Cincinnati Model Agency
6047 Montgomery Rd.
Cincinnati, OH 45213

Ford Models & Talent
1300 E. 9th St., Ste. 1640
Cleveland, OH 44114

S2 Management Group
844 N. High St.
Columbus, OH 43215

Creative Talent Company
5864 Nike Dr.
Hilliard, OH 43026

Z Model Management
985 Mediterranean Ave.
Columbus, OH 43229

The Right Direction Model Mgmt.
4770 Indianola Ave., Ste. 160
Columbus, OH 43214

Wings Model Management
906 Main St., Ste. 207
Cincinnati, OH 45202

Bowman Agency
PO Box 4071
Lancaster, PA 17604

Click Models
216 Green Tree St.
West Chester, PA 19382

Donatelli Model & Casting Agency
156 Madison Ave.
Reading, PA 19605

Expressions Agency
220 Church St.
Philadelphia, PA 19106

Fashion Mystique Modeling Agency
611 N. Mountain Rd.
Harrisburg, PA 17112

Docherty
109 Market St., 3rd FL
Pittsburg, PA 15222

Main Line Models
1215 W. Baltimore Pike, Ste 9
Media, PA 19063

Marilyn E. Kane Model Mgmt
1022 N. Main St.
Butler, PA 16001

Millenia Model & Talent Management
4902 Carlisle Pike, Ste 228
Mechanicsburg, PA 17055

Millennium Model Management
601 S. Henderson Rd., Ste. 203
King of Prussia, PA 19406

MMA – Model Management Agency
106 S. Bellevue Ave., Ste. 205

Plaza 7 Model & Talent Reps
160 North Gulph Rd.
King of Prussia, PA 17406

Prestige Modeling
10028 Frankstown Rd.
Pittsburg, PA 15235

Pro-Model Management
1244 Hamilton St.
Allentown, PA 18102

Reinhard Model & Talent Agency
2021 Arch St.
Philadelphia, PA 19103

Slickis Models
1777 Walton Rd., Ste. 204
Blue Bell, PA 19422

The Talent Group, Inc.
2820 Smallman St.
Pittsburg, PA 15222

Vision Model & Talent
14999 Colver Dr.
W. Springfield, PA 16443

Vision Model Management
5120 West Ridge Rd.
Erie, PA 16506

ASA International Model & Talent
3926 Wesley St., Ste 702
Myrtle Beach, SC 29577

Betty Lane Models Talent Agency
951 Doyle St.
Orangeburg, SC 29115

Carolina Winds Productions
141 Gadsden St.
Chester, SC 29706

Dreams Unlimited
959 Maudin Rd., Ste. 105
Greenville, SC 29607

Harvey Talent
PO Box 50505
Columbia, SC 29250

Millie Lewis of Columbia Inc.
3612 Landmark Dr., Ste. D
Columbia, SC 29204

Millie Lewis Models & Talent
1904 Savannah Highway
Charleston, SC 29407

Showcase Models & Talent
1200 33rd Ave. South
N> Myrtle Beach, SC 29582

Advantage Models & Talent
4825 Trousdale Dr., Ste. 211
Nashville, TN 37220

AMAX-Models, Actors, Extras
4121 Hillsboro Rd., Ste 300
Nashville, TN 37215

Ambiance Models & Talent
1096 Dayton Blvd.
Chattanooga, TN 37405

Billy Deaton Talent
1300 Division St., Ste 102
Nashville, TN 37203

The Cannon Group
1719 Division St., Ste. 417W
Nashville, TN 37203

Capitol Management & Talent Group
1300 Division St., Ste 200
Nashville, TN 37203

Career Model & Talent Management
PO Box 977
Hendersonville, TN 37077

Coleman Model & Talent Agency
417 E. Iris Dr.
Nashville, TN 37204

Colors Agency, Inc.
408 S. Front St.
Memphis, TN 38103

Jo-Susan Model & Talent
2817 W. End Ave., Ste. 210
Nashville, TN 37203

Moore/Higgs Casting
417 Welshwood Dr.
Nashville, TN 37211

Richmond Casting Group
1719 Westend Ave.
Nashville, TN 37203

Robbins Model & Talent
176 Walnut Bend
Memphis, TN 38018

The Donna Groff Agency Inc.
PO Box 382517
Memphis, TN 38183-2517

The Hurd Agency
500 Eversholt Ct.
Antioch, TN 37013

The Harper/Spear Agency
PO Box 158779
Nashville, TN 37215

TML Talent Agency
PO Box 40763
Nashville, TN 37204

Avant Models & Casting, Inc.
85 NE Loop 410, Ste. 219
San Antonio, TX 78216

Calliope Talent & Modeling Agency
1802 NE Loop 410, Ste. 107
San Antonio, TX 78217

Clipse Model & Talent Management
1420 W. Mockingbird Lane, Ste. 280
Dallas, TX 75247

Condra/Artist Model & Talent
13300 Old Blanco Rd., Ste. 201
San Antonio, TX 78216

DB Talent
3107 Slaughter Lane W.
Austin, TX 78748-5705

Diane Dick Int. Modeling & Talent
1410 S. Washington St.
Amarillo, TX 79102

First Models & Talent Agency
5433 Westheimer Ave., Ste. 310
Houston, TX 77056

International Model & Talent Agency
2727 Kirby Dr.
Houston, TX 77098

Kim Dawson Agency Inc-Talent Div.
2710 N. Stemmons Freeway, Ste.
Dallas, TX 75207

Mary Collins Agency
2909 Cole Ave., Ste. 250
Dallas, TX 75204

Pastorini-Bosby Talent Agency
3013 Fountainview Dr., Ste. 240
Houston, TX 77057-6120

Reflections Fashion & Talent Agency
110 Alta Plaza
Corpus Christi, TX 78411

Sherry Young/Mad Hatter
2620 Fountainview, Ste. 212
Houston, TX 77057

Talents International
9449 Briar Forest, Ste. 3105
Houston, TX 77063

The Talent House
812 N. Virginia St., Ste. 204
El Paso, TX 79902

The William Ware Agency
2931 Irving Blvd.
Dallas, TX 75247

Tomas Agency
14275 Midway Rd., Ste. 220
Addison, TX 75001

Neal Hamil Agency
7887 San Felipe Ste. 227
Houston, TX 77063

The Campbell Agency
3906 Lemmon Ave., Ste. 200
Dallas, TX 75219

20 South Talent Agency
114 Goodman St.
Charlottesville, VA 22902

Beaute
10405 Huneen Station Rd.
Vienna, VA 22181

Beautiful People Model and Talent
3237 Tyreneck Rd.
Portsmouth, VA 23703

Elan Models and Artists
1200 N. Veitch St., Ste. 1
Arlington, VA 22201

Erickson Casting
8201 Greensboro Dr., Ste. 214
McLean, VA 22102

Encore!
17 Keith's Lane
Alexandria, VA 22314

Evie Mansfield Modeling Agency
505 S. Independence Blvd., Ste. 205
Virginia Beach, VA 23452

Glamour Modeling & Talent
1115 Independence Blvd., Ste 110
Virginia Beach, VA 23455

Marilyn's Professional Model & Talent
2721 Charstone Dr.
Midlothian, VA 23113

Millennium Model & Talent Mgmt.
8321 Old Courthouse Rd.
Vienna, VA 22182

Model and Talent Management
249 S. Van Dorn St.
Alexandria, VA 22304

New Faces Models
8230 Leesburg Pike, Ste. 520
Vienna, VA 22182

On Call Models & Talent
946 Ferryman Quay
Chesapeake, VA 23323

Steinhart Norton Talent Mgmt.
312 Arcpic Crescent
Virginia Beach, VA 23451

Studio M
1719 Wainwright Dr.
Reston, VA 20190

Anne Schwab's Model Store
906 D St. NE
Washington, DC 20002

Central Casting Incorporated
623 Pennsylvania Ave. SE
Washington, DC 20003

Doran Model & Talent Agency
1301 Massachusetts Ave. NW, Ste. 701
Washington,, DC 20005

MULTICULTURAL MODELING & TALENT AGENCIES

Black Models U.S.A
Black Models U.S.A.:Black Women and Females:Life Styles:Casino Employment:Black Entertainment:African American:Modeling Agency:Talent Agency:Actors and Actress:Models News and Media:Open Calls:Adult Entertainment:Employment and Jobs:New York City...
www.musecube.com

Black Models U.S.A.©1998
Black Models U.S.A.©1998 Modeling Talent Agency-Hosting. In Business 7 Yrs. We Franchise Modeling Agencies. Represent and Manage:Minors 1-12, Teens 13-17, Adults 18-60, Models, Singers, Actors, Dancers, Plus Size Models, Hip Hop, Rap, Black Music...
www.blackmodelusa.com

Black Entertainment Television Showcase©2004
Black Models U.S.A. Television©2004 Hip Hop Rap Music Video Fashion Models Magazines Websites Providing free! Services Resourses and Directory For Unsigned Models Dancers Actors Singers Spoken Word Artists Exotic Dancers Sports and Athletic...
blackentertainmenttelevisionshow.cjb.net

Setworkers.Com Arts And Entertainment Jobs
Arts & Entertainment Crew Jobs For Tv & Film Jobs 100"s Entertainment Jobs Set & Stage Workers Entry Level Apply To Entertainment Film Crew Movie Studios Job Listings, Resume Service. Film Production Entertainment Film Crew Job Employment Opportunities....
www.setworkers.com

Black Models U.S.A. Events : Models Open Call
Black Models U.S.A. Is A Black Owned Modeling and Talent Agency Website Based In New Haven Connecticut offers its members and online visitors free! Websites Urban Perspective. The Website Information On hip hop rap black music black entertainment...
www.urbanjoint.com

Black Models USA©1998 - New York City
Black Models U.S.A.©1998 Modeling and Talent Agency. Fashion Models Magazine, Weekly Television Show. Free! Models Websites w/Photos Flash and Music, Chatrooms, Forums, Message Boards, Photo Albums...
blackmodelsusa.bravehost.com

Black Casting - Mis-Education of Corrine Thomas - New Haven
Entertainment performers including Actors, Models, Dancers, & Vocalist can use Star Search Casting's talent database to market their resume & profile to Modeling Agencies, Agents, Talent Agencies, Casting Directors, production companies...
starsearchcasting.com

Willie Habersham Williams Jr
Models Modeling and Modeling Agencies > Authors Writers and Journalists > Black Hall of Fame > Actors and Actress > Black Music Awards > Entertainment > Dance > Film Movies and Videos > Arts and Entertainment >Performing...
www.williewilliamsjr.cjb.net

Black Models U.S.A
Reginal > North America > United States > New York City > Localities > N > Arts and Entertainment > Performing Arts > Fashion Models > Modeling Agencies > Modeling > Models > Black > Black Models > African American > Ethnic...
www.blackmodelsusa.tk

Talent Agency Troupers.Com. Actors Actresses Models For Film
Talent Agent For Actors And Actresses. Agency Representation For Modeling, Acting, Television And Film Work. All Entertainers Represented Worldwide. Singing, Dancing, Modeling, Acting....
www.troupers.com

Entertainment Jobs And Internships - Entertainment Careers.Net
Entertainment Jobs In Film Television Music Animation Studios Networks: Disney, Fox, New Line, Paramount, Sony, Universal, Warner Bros. More....
www.entertainmentcareers.net

Entertainment Jobs & Internships - Entertainment Careers.Net
Entertainment Jobs In Film Television Music Animation Studios Networks: Disney, Fox, New Line, Paramount, Sony, Universal, Warner Bros. More....
www.entertainment-jobs.com

African League News©2004
A Black Organization Controlled By Black People. News and Media > Black Music > Black Entertainment > Arts and Entertainment > Employment and Jobs > Hip Hop > Rap > Models and Modeling > Organizations > Non Profits...
africanleaguepress.cjb.net

Ethnicspotlight.com
Ethnic Talent agency, Black talent, Asian Talent, London based, offering Ethnic singers, dancers, models, actors, actresses the channce to join and sign up to get work in the TV, music and entertainment industry...
www.ethnicspotlight.com

Black Beauty
DJ's, Models, Male & Female Exotic Dancers, Adult Film Stars - Black Beauty Enterprises, Inc. For the Best in Black...
www.blackbeautyenterprises.com

Model Magazine, LLC
Talent agency. Modeling agency. Publishing services. Job search for talented performers. Talent management. Talent scout. Models, actors, performers wanted...
www.modelmagazine.net

Modeling & Talent Agency In Washington, Dc Books Top Talent!
Anne Schwab'S Model Store Is A Washington, Dc Modeling And Talent Agency Booking Accomplished Print Models, Fashion Models, Actors & Narrators....
www.themodelstore.com

Black Models U.S.A. - Models and Modeling Directory - Connec
Black Models U.S.A. Models and Modeling Directory Connecticut. Business > Arts and Entertainment > Agents Agencies > Websites Hosting > Portfolio Hosting > Models Pages > Free > Free Websites > Models Modeling...
blackmodelusa.instantdirectory.net

Fla Talent Splash
Florida, Talent, Models, Casting, Cast, Calls, Modeling, Actors, Resumes, Production, Camera, Edit, Video, Film, Mm, Shoot, Shot, Movie, Stars, Photo, Photography, Film, Betacam, 70mm, Acting, Theater, Stage, Sound, Lighting, Recording, Music, Musician...
www.flatalent.com

Teen Modeling Website MyModelFinder
Become a model at MyModelFinder Premier online model site features Models portfolios and teen models, Adult models, Preteen Models and child models. Hispanic teen modeling website agent and agency directory...
www.mymodelfinder.com

Black Models U.S.A. : Agency : New Haven CT US
Modeling Agencies > United States > North America > N > Fashion Models > Modeling > Models > Free Portfolio Hosting For Models, Actors, (TFP) Photo Shoots, Photo Albums, Photographers, Modeling Agencies, Talent Agencies, Schools...
www.onetalentsource.com

Models And Talent.Com: Modeling Agencies, Modeling Portfolios
Casting & Scouting Service For Models, Actors, Agents & Casting Directors. Online Model Portfolios, Actors Headshots, Advice, Guidance And Legitimate Opportunities For Models And Talent....
modelsandtalent.com

Pro-model & Talent Management - Ohio"s Premiere Modeling And
Pro-model & Talent Management Is Not Only A Mother Agency With Models And Talent Working In New York, Los Angeles, Miami, And Europe, It Is Also The Largest Commercial Agency In Ohio With Models And Talent Working Throughout The State, Into Pennsylvania...
www.promodelmgmt.com

Black Models U.S.A. Models Message Boards and Forums
Models Message Boards and Forums > United States > North America > American Top Models > Television > Fashion Shows > Runway Models > Models Contests > Models Pageants > Modeling Agencies > Modeling Agency > Ethnic Models...
runmodels.invisionplus.net

Jabo Modeling Management Agency: Modeling Agency For Preteen
Modeling Agency Representing Talent With Models Camp, Runway Shows, Pageants, Open Calls, Modeling Evaluation, Conventions And Fashion Wear - Preteen, Teen, Young Adult Male And Female Models. Introducing Jabo Models "Pressbook" Portfolios! Become A...
www.jabomm.com

Australian Model Directory - models, modelling, modeling
Australian Model Directory - models, modelling, modeling, fashion models, modelling agencies, modeling agencies, model photographers, model search, supermodels, model portfolios, model management, girls, woman, women, beauty, model photos, female models...
www.modeldirectory.com.au

MyModelFinder
Modeling and Talent Community features online portfolio and modeling resources to talented teen and adult models.. MyModelFinder helping models get the exposure you deserve.The number one modeling site MyModelFinder...
mymodelfinder.com

Film and Television Production Companies
Black Models USA Television Show CTV Channel 27. Executive Vice President and Co-Owner, Internships, Grants, Scharlorships, Loans, Students, Trainees, and Volunteers. Models, Production Assts.Television Anchors, Talk Show Hosts, News Reporters, Girl...
www.newsandmedia.cjb.net

DeMar P.R. - Where talent transcends beauty
Susanne DeMar: Saint Louis modeling agency featuring model training, photo shoots, acting classes, talent agency, fashion jobs, models portfolio.Children, Teens, Men, Women, DJ's Voice Talent, Hair dressers, make up artists are all welcome...
www.demarpr.com

Agencies: Black Models U.S.A
Black Models U.S.A. Modeling Agency From New Haven, U.S.A. Presented On Modeling Magazine jurgita.Com. Agency Specializes In Runway, Print, Television, Showroom, Fitness, Ethnic, Models Over 30+ Catalog, Editorial, Runway, Body Part, Showroom...
www.jurgita.com

XTC Models
XTC Models Talent Management, Adult Modeling, Models, Locations, Photographers, videographers, Talent...
www.xtc-models.com

Boca Models
Models: Model, Modeling Agency Located In South Florida, Fl. We Serve Clients All Over The World. We Are A Full-service Modeling Agency With Models For Print, Fashion, Tv, Commercials, Film & More....
www.bocamodels.com

MyModelFinder
Modeling Website MyModelFinder Features portfolios of aspiring Teen models, Child models, adult models and Hispanic teen models. Become a model at the premier online modeling site www.mymodelfinder.com...
www.mymodelfinder.info

Elegant Models - Your best choice for modeling talent
We provide modeling talent for your video, print, runway, and event modeling needs. We're always looking for new talent...
www.elegantmodels.net

Modeling Agencies In America©2004
Black Models U.S.A.©1998 230 Grand Avenue-Suite # 576 New Haven, CT 06513. Child Models Ages 12-17. Free! Models Websites w/ Your Name. Directory Guide Index. Talent Booking Assignments Pay $$$. Rules Laws Regulations. Who's Who In Modeling...
modelingagenciesinamerica.cjb.net

Black Models U.S.A. Online Magazine :: Index
Black Models U.S.A. Online Magazine Models Modeling Modeling Agencies Fashion Shows Runway Models Modeling Events Open Calls Auditions Employment Magazines Fashion and Fashion Designers Models Websites Models Portfolios Headshots Photo Shoots...
www.forums.com

Black Models Pictures
Welcome To My Modeling Portfolio. My Name Is Karen. I"m Available For Acting, Modeling And To Take Pictures For Professional Photographers. My Profile Hi, My Name Is Karen, I"m A 20 Years Old Black Female Model And Also An Aspiring Actress. My Modeling...
www.portfolioresume.8k.com

Black Models U.S.A.©1998 Directory and Search Engine
Black Models U.S.A.©1998 Models Directory and Search Engine: Ethnic > Black > African American > Lesbian > Lesbians > Femme Women > Employment > Jobs > Administrative > Clerical > Retail > Models > Modeling > Fashion Beauty > Agency > Agencies...
blackmodelsusa.instantdirectory.net

Quest Model Management | Singapore
Quest Models, Quest Model Management, Quest, Q, Management, Model, Models, Talent, Talents, Choreography, Agency, Modelling Agency, Modeling Agency, Singapore, Styling, Catwalk, Fashion, Runway, Glamour, Film, Consultancy, Mannequin Studio, Mannequin...
www.quest-models.com

Modeling Camp Home Page
Modeling Agency, Modeling Camp, Modeling, Fashion, Models, Modeling Agency, Model Camp, Model Work, Runway, Fashion Show, Fashion And Models, Model Fashion, Model, Make-up, Modeling In Va, Va, Virginia, Models In Va, Summer Camp, Camp, Model Camp...
www.modelingcamp.com

Productions Plus - A National Modeling Talent And Casting Agency
Productions Plus - A National Modeling Talent And Casting Agency Productions Plus - A National Talent And Casting Agency For Auto Shows, Conventions, Trade Shows And Special Events Needing A Model, Or Models For Modeling, Talent, Promotions, Narrators...
www.productions-plus.com

Model Venue
Model Venue- We have all types of models Fashion models model agency model agencies female models male models fitness models florida new york california la dallas texas chicago modeling runway georgia swimsuit acting actor arizona tennessee...
www.modelvenue.com

Delightful Models
Modeling website with extensive picture galleries of sexy models, import models, model intervies, modeling forums and monthly modeling contest...
www.delightfulmodels.com

America Models.Com America Models Networks: Modeling Resource
America Models Provides Aspiring Models With A Means To Promote Themselves, Via Electronic Compcards And Online Modeling Portfolios, Our Free Membership Allows Models To Index One Thumbnail Photo For Inclusion On Our Website, Model Agencies, Modeling...
www.americamodels.com

Modeling At Goldtan Models
Models Directory Spotlight Model Modeling Tips Auditions Models Store Need To Hire A Model...
models.rarmall.com

MyModelFinder
A Modeling and Talent community featuring online portfolios.MyModelFinder.Com the number 1 online international modeling site Features Teen Models super models Male Models female models Runway Modeling...
www.mymodelfinder.net

Jobs in film and television
Jobs in Film, TV, Radio and Broadcasting. We'll show you how to find jobs in the entertainment industry...
johnnyblad51 - Models Modeling
Stop Dreaming About A Career In Modeling And Talent. Models Modeling Inc. Offers The Beginner And Professional Alike Your Own Web Site Where You Can Be Discovered And Get Noticed! A Team Of Professional Models And Photographers Stand Ready To Act As Your...
www.modelsmodeling.com

V3 Models
Southern California Model & talent management. Male and Female models. View portfolio of models and talent, fashion, bikini models, promotional models and more...
www.v3models.com

The Lyons Group Home
Modeling, Models, Model, Modeling Agency, New York, Sports Models, Fashion Models, Body Models, Kids Models, Actors, Acting Agency, Curves, Plus Size Models, Big And Tall...
www.lyonsgroupny.com

Affinity Model And Talent
Model, Modeling, Modelling, Models, Model Agency, Cast, Casting, Casting...
www.affinitytalent.com

welcome to Model Material
Model scouts for all modeling agency's start your modeling career with online profile new models wanted tip on how to get started as a model in England new models wanted...
www.modelmaterial.netfirms.com

Stone Model & Talent Agency - Ohios #1 For Fashion Modeling
Modeling Agency, Photographs, Models...
www.stonemodels.com

ActingModeling.com - actors, actresses, models, auditions
About Acting, acting, actor, actress, actors, Actor, Actos, Actress, actresses! Find all acting agency, act, talent, acting biz, acting agency, model, modeling, models, Model, Models, supermodels, fashion, fashions, art, actor resources, talent agent...
www.actingmodeling.com

Emerald Entertainment
Emerald Coast Talent Directory, North West Florida, Models, Actors, Singers, Dancers, Talent Search...
www.emeralddirectory.com

Surfview Entertainment : Film And Movie Finance, Production
Helps Independent Producer Find Film Finance For Their Movie, Television And Film Productions....
www.surfview.com

PROACTOR TOOLS
Connection for employment opportunities and jobs in the entertainment industry across the country. Need Acting Auditions? Modeling Agency Talent Search Baby Modeling and Actors of All Ages. Reach all entertainment industry professionals quickly & cost e...
www.proactortools.com

Job Modeling Information to Help You be a Successful Model
Job modeling information about helpful modeling topics such as: beauty, modeling agencies, requirements, make up, nail care, exercise advice, hair types, modeling terms, runway, and posture and poise...
www.agency-modeling-look.com

Tucson Model Group - Model, Talent, Acting Search Tucson Arizona
Tucson Model Group Helps Aspiring Models, Actors And Actresses, And Performing Artists In Tucson Arizona Break Into The Modeling, Talent And Acting Industries....
www.tucsonmodelgroup.com

Black Models USA Online Magazine
Black Models U.S.A. Online MagazineÂ©1998 Add Your News Stories Events and Photos free!. Your Image and Name Is Everything-Modeling Agencies Work For You 24/7. Black Models Fashion Shows News First 1st. Fridays Arts Entertainment Music Fashion Events...
black.iuplog.com

Hollywood North | Acting | Modeling | Movies | Television
Adult Hollywood North - Where Actors, Models, Casting Directors, Agents, Producers and the Entertainment Industry meet...
www.adulthollywoodnorth.com

Armenian Models And Fashion Industry Online
Models.Am Designs Custom Web Sites For Modeling Agencies And Features Models, Modeling Agencies, Model Agency Web Sites, Scouting Events, Models Online, Model Databases, New Models....
www.models.am

Lemmecheck Entertainment : Glamour Models, Supermodels And
Beautiful Ladies, Models And Supermodels From Non-nude To Glamour Plus A Searchable Resource For The Modeling, Talent, Celebrity And Entertainment Industries.....Your Gateway To Beauty!...
www.the-hitlist.com

Place Model Management : Welcome !
Place, Place Model Management, Placemodels, Placemodels, Modelagency, Modelagenturen, Model Agency, Modelagentur, Modeling Agency, Place Models, Mode, Model, Model, Modelle, Models, Men, Women, Girls, Boys, MäDchen, Jungen, Modeling, Hair, Makeup, Make...
www.placemodels.de

Impact Models and Entertainment
Provides models and talent for print, fashion, conventions, film and events....
www.impact-models.com

Black Models U.S.A. Modeling Agency
Black Models USA Modeling Agency Website Creations-Hosting Photo Studio. Free Membership 485 Registered Members: Chatrooms-Websites-Portfolios-Forums-Message Boards-Guestbooks-Email. Representing Models Male-Female Age 13-50...
www.blackmodel.tk

Welcome To The Talent Shop Website. Please Be Our Guest And Look
The Talent Shop Has Been Casting For Producers, Photographers, And Directors For 20 Years. We Provide Talent In All Fields Including, Male Models, Female Models, Children Models For Film, Voice, And Broadcast....
www.thetalentshopgroup.com

Modeling Industry News - Who's Who In New Haven Connecticut
Modeling Industry News : Who's Who : New Haven : Connecticut : Willie Williams Jr : Modeling Consultant : Contract Manager : Models and Modeling : Arts and Entertainment : Runway Models : Modeling Agency : free! Models Websites...
williewilliamsjr.freelancebank.com

Caryn International Model And Talent Center
Caryn International Is The Leading Modeling And Talent Agency In The Midwest. We Develop Models And Actors For Local, National And International Markets. We Have Represented Over 5, 000 Models And Actors And Have A Corporate Client Base Of Over 2, 000...
www.caryninternational.com

Angels Unlimited Modeling Center
The Models Survival Guide And How To Survive And Prosper In The World Of Modeling With Information On Modeling Contests, Bikini Contests , Model Promotions, Modeling Agencies, Modeling Resources, Modeling Career Information, And Modeling Facts, Fiction...
www.trimbodies.net

Welcome To Ajente.Com - The Source For Urban Entertainment
Black Entertainment, Entertainment, Blacks, African Americans, Black Singles, Black Talent, Celebrities, Personals, Black Love, Events, Concerts, Travel, Prizes, Black Music, Music, R&B, Hip Hop, Rap, Jazz, Reggae, Gospel, Blues, Black Movies, Movies, Black Vhs, Black Dvd, Black...
www.nwblackent.com

Eight Model & Talent
Eight Models, Talent, Promotions, Models, Promotional Models, Hartford, Ct, New York, Ny, New Jersey, Nj, Pennsylvania, Pa, California, Ca...
www.eightmt.com

Professional Models Looking For Modeling Jobs
Offers amateur and professional women, teen, child and preteen models looking for modeling jobs legitimate promotional websites...
www.1st-in-models-modeling.com

Mitch Karam Casablanca International Model & Talent Agency
Models, Artist, Talent, Tv, Movie, Agent, Agency, Photographer, Nashville, Tn, Management, Actress...
www.mitchkaram.com

Moviework Now Casting - The Source For Arizona Talent!
Moviework Now Casting Is Needing Actors, Models, And Movie Extras For Several Upcoming Arizona Feature Film, Television, And Commercial Projects....
www.movieworknow.com

Model Madness
Model Madness is a 3-day modeling and acting convention held in Washington, DC. Model Madness is an opportunity for models and actors to network with the Nation's Top Agents, Scouts, Casting Directors, Talent Agents and Photographers at one major event...
www.modelmadness.com

Adult Acting | Adult Modeling | Adult Hollywood North | Movies
Hollywood North - Where Actors, Models, Casting Directors, Agents, Producers and the Entertainment Industry meet...
www.adulthollywood-north.com

Actors, Models, Talent | Print, Television, Film | Green Agency
Agency For Women, Men And Children Models...
www.greenagency.com

Louisiana Modeling Agency: Michael Turney Modeling Agency.
Louisiana Modeling Agency: Michael Turney Modeling Agency. Professional Modeling For Male Models, Female Models & Child Models. Specializing In Runway Modeling, Portfolios, Pageant Training & Model Training....
www.michaelturneyagency.com

Ariza Inc.
Talent and modeling agency representing talented child, male, female models and actors for commercials, films and television....
www.arizatalent.com

mysitetitle
Model Modeling Actress Acting Fitness Trainer Promotion Movie Television Advertising...
heatherstorm.com

Las Olas Models And Talent Agency Fort Lauderdale - Miami
Las Olas Models And Talent Agency Fort Lauderdale - Miami Florida Women Men Kids Teens Actors Actress Tv Print Film Runway W Om En ©2000 Las Olas Models & Talent. All Rights Reserved. Model And Talent Agency Ft. Lauderdale - Miami Florida Miami Florida...
lasolasmodels.com

Florida Beach Guy Models finds jobs for female fashion models
Florida Beach Guy Models finds jobs for female fashion models, female adult entertainment models, male models, and voice over actors.We market our models world wide...
www.floridabeachguymodels.com

Www.Allseasonspc.Com
Modeling Jobs Now Available Nationwide Aspiring/Current Child, Teen And Adult Models Wanted Now For Print, Fashion, Plus-size, Youth, Promotional, And Sophisticated Modeling Bookings Available Through Top Agencies Nationally. Www.Fashionrocknow.Com...
www.allseasonspc.com

Www.Gowellgames.Com
Modeling Jobs Now Available Nationwide Aspiring/Current Child, Teen And Adult Models Wanted Now For Print, Fashion, Plus-size, Youth, Promotional, And Sophisticated Modeling Bookings Available Through Top Agencies Nationally. Www.Fashionrocknow.Com...
www.gowellgames.com

Www.Ojibwaynativeart.Com
Modeling Jobs Now Available Nationwide Aspiring/Current Child, Teen And Adult Models Wanted Now For Print, Fashion, Plus-size, Youth, Promotional, And Sophisticated Modeling Bookings Available Through Top Agencies Nationally. Www.Fashionrocknow.Com...
www.ojibwaynativeart.com

Index Of Mailing Label Listing:

***There is no promise or guarantee that all of these Casting Director/Agents are still working with the individual companies, or if still in business.**

ATLANTA CASTING DIRECTORS & AGENTS

Dara Dearborn
Code Talent Mgmt Inc.
2849 Piedmont Rd. NE #1
Atlanta, GA 30305

Teresa Kellar/Stephanie Duke
The Madison Agency
PO Box 46717
Atlanta, GA 31146

Ted Borden
Ted Borden & Associates LTD
2434 Adina Drive NE, Suite B
Atlanta, GA 30324

Brenda Pauley
People Store
2004 Rockledge Rd NE
Atlanta, GA 30324

Kris Redding
Kris Redding Casting
PO Box 55257
Atlanta, GA 30308

Kelly Kelly Enterprises Inc
10945 State Bridge
Suite 401-316
Alpharetta, GA 30022

Mystie Buice/Gail Houghton
Houghton Talent Inc
919 Collier Rd NW
Atlanta, GA 30318

John Templeton
Tara Talent Agency
696 Mt. Zion Rd. #C1
Jonesboro, GA 30296

Randall Edwards/Kathy Hardegree
Atlanta Models & Talent
2970 Peachtree Rd. NW #660
Atlanta, GA 30305

Darlene McCrary
Kennedy Model & Talent, Glyn
16 Willow Bend Dr.
Cartersville, GA 30121

Joan Taibbi
TMA Talents
1702 Dunwoody Pl NE
Atlanta, GA 30324

Kitty Bundy
Elite Atlanta
1708 Peachtree NW #210
Atlanta, GA 30309

Donna Summers
Voice Casting Net work
8950 Laurel Way, Suite 200
Alpharetta, GA 30202

Connie Suhr/Extras
Connie SUHR
PO Box 190424
Atlanta, GA 31119

Rona Burns
The Burns Agency
602 Hammett Drive
Decatur, GA 30032

Richard Hutchinson
Richard S Hutchinson Mgmt Inc
3071 Arden Rd NW
Atlanta, GA 30305

Mark Cook
L'Agence Suite 60
5901-C Peachtree Dunwoody Rd.
Atlanta, GA 30328

Jen Kelley
People Store
2004 Rockledge Rd. NE
Atlanta, GA 30324

Julie Martin/Lynnita Terrell
Arlene Wilson Model Mgmt
887 W. Marietta St. #N101
Atlanta, GA 30318

Joy Pervis
Hot Shot Kids/Teens
2004 Rockledge Rd NE
Atlanta, GA 30324

Kimberly Holt
Tara Talent Agency
696 Mt. Zion Rd. #C-1
Jonesboro, GA 30236

Don Slaton
Don Slaton Casting
421 Kirk Rd
Decatur, GA 30030

Christy Clark/Suzanne Lear
Genesis Modes & Talent
1465 Northside Dr. #120
Atlanta, GA 30309

Brenda Pauley
The Talent Group
561 W Pike St. #100A
Lawrenceville, GA 30345

AL, NC, SC, TN REGIONAL AGENTS

Sherry Graves/Michael Fulmer
Real People Models &
Talent/Kiddin Around
1827 First Avenue North #209
Birmingham, AL 35203

Beverly Brock
The Brock Agency
329 13th Avenue NW
Hickory, NC 28601

Jackie Pressley
Capital Artists, Inc
1405 Bloomingdale Drive
Cary, NC 27511

William Pettit
William Pettit Agency
Box 11798
Charlotte, NC 28220

Rona Burns
The Burns Agency
1255 Marlborough Lane
Winston-Salem, NC 27105

Marilyn Green/Kathie Bozard
Marilyn's Inc
601 Norwalk St.
Greensboro, NC 27401

Jackie Pressley
CAI Kids
1405 Bloomingdale Drive
Cary, NC 27511

Betty Cark
TML Talent Agency
PO Box 40763
Nashville, TN 37204

Ashley Russell
Amax Models & Talent Agency
2815 Bransford Avenue
Nashville, TN 37204

Sharon Smith
Talent Trek Agency Nashville
2021 21st Avenue South #102
Nashville, TN 37212

Lou Crisuolo/Mary James Crisuolo
MJM Talent Representative
2011 Carolina Beach Road
Wilmington, NC 28401

Terri Alexander
Sass Talent Agency
8419 Clear Meadow Lane
Charlotte, NC 28227

Sharon Dunn
Carolina Talent
1230 W Morehead St #110
Charlotte, NC 28208

Debby Lee
Ice Model/Talent Management
McMullen Creek Market
8318 Pineville-Matthews Road
#265
Charlotte. NC 28226

Anne Swindell
The Talent Connection
338 N. Elm St. #204
Greensboro, NC 27401

K. Michelle/J. O'Neal/L. Newcomb
JTA, INC
820 EAST BOULEVARD
CHARLOTTE, NC 28203

Stephanie Williams/Betsy Briggs-Cathart
Dan Agency Models & Talent
209 10th Avenue South #301
Nashville, TN 37203

Betty Clark
TML Talent Agency
PO Box 40763
Nashville, TN 37204

Annette Amato Outlan/Jo Wagoner Bracey
Colors Talent Agency
1985 Madison #3
Memphis, TN 38104

Tammy Hopkins
Actors and Entertainment Inc
PO Box 486
Keeseville, NC 27285

Tommi Long Jones
Tommi Long Jones Talent
320 Settlers Lane
Charlotte, NC 28202

Russell Wiggs
Artist's Resource Agency
174 Hunters Glen Drive
Summerfield, NC 27358

Mandy Silla
Carolina Talent
1230 W Morehead St #110
Charlotte, NC 28208

Juanell Walker
Talent Trek Agency Asheville
825-C Merrimon Ave. #356
Asheville, NC 28804

TENNESSEE AGENTS

Jimmi McCarter
The Cannon Group
1622 16th Avenue S #500
Nashville, TN 37212

Misty Griggs
Advantage Models & Talent
230 Franklin Road #802
Franklin, TN 37064

Tammy Hopkins
Talent Trek Agency
540 Kingston Pike #450
Knoxville, TN 37919

Karilyn Allers
Career Model & Talent Mgmt.
PO Box 977
Hendersonville, TN 37077

Charlie Peterson
Harvest Talent
2800 Bush River Road #3B
Columbia, SC 29210

Dee Voight
Dee's Casting
160 East Jane Circle
Hampton, GA 30228

Kim Petrosky
Kim Petrosky Casting
3204 West End Circle #2
Nashville, TN 37203

Marie Prater
Prater & Prater, LLC
2642 O'Neal Circle
Birmingham, AL 35226

Lyn Richmond
Richmond Casting Group
1719 West End Ave. #306 West
Nashville, TN 37203

Mitzi Corrigan/Paige Johnston
Corrigan & Johnson Casting
3006 N. Davidson St.
Charlotte, NC 28205

Lori Nelson
Nelson-Miller Casting
395 Promenade Ct.
Marietta, GA 30064

CASTING DIRECTORS CHICAGO

Terrance A. Hurd
The Hurd Agency
500 Eversholt Ct.
Antioch, TN 37013

Shay Bentley Griffin
Chez Casting
2221 Peachtree Rd. NE X-14
Atlanta, GA 30309

Craig Fincannon/Mark Fincannon
Fincannon & Associates
1235 N 23rd St.
Wilmington, NC 28405

Annette Stillwell
Annette Stillwell Casting
PO Box 53017
Atlanta, GA 30355

Cher Payne
Cherra Casting
5888 Third Cove #8
Memphis, TN 38134

Regina Moore
Moore Casting
417 Welshwood Drive #109
Nashville, TN 37211

Jo Ann Smith
Atlanta Casting Inc.
650 Mt. Zion Road #C1
Jonesboro, GA 30236

R. Futch/M. Mullins
Futch & Mullins Casting
PO Box 21144
Charleston, SC 29413

Jane Alderman
Jane Alderman Casting
Act One Studios
640 North LaSalle #535
Chicago, IL 60610

AL, NC, SC, TN, GA REGIONAL AGENTS

Shirley Crumley
Shirley Crumley
1825 Molly Jane
Birmingham, AL 35235

Jim Kup
Jimmy Kup Casting
Box 22748
Nashville, TN 37202

K. Gallagher/M. Sill/s. Baskin
Carolina Talent
PO Box 36422
Charlotte, NC 28236

John King
John King Productions
244 Madison Ave. #393
New York, NY 10016

Tammy Hopkins
A&E (Actors & Entertainers)
PO Box 486
Kernersville, NC 28285

Jo Doster
Jo Doster
PO Box 120641
Nashville, TN 37212

Don Slaton
Don Slaton Casting
421 Kirk Road
Decatur, GA 30030

Chicago Model Search Inc
360 West Illinois
Chicago, IL 60601

Bass Casting
340 North Ogden
Chicago, IL 60607

Claire Simon
Simon Casting
1512 North Fremont #202
Chicago, IL 60607

Mindy Verson
Chicago Radio Casting
232 East Ohio St.
Chicago, IL 60657

Norman Schucart
Norman Schucart Ent.
1417 Green Bay Road
Highland Park, IL 60035

Tracey Tarantino
Zzazz Productions
113 Covington Ct
Oak Brook, IL 60523

Hal Watkins/Rik Kristinat
Hollyrik & Heitz Casting
920 N. Franklin, Suite 205
Chicago, IL 60610

Casting Dept.
3rd Coast Artists
641 W. Lake St Suite 402
Chicago, IL 60661

CHICAGO ACTORS & MODELS AGENTS

Lily Ho
Lily's Talent Agency
1301 W. Washington #B
Chicago, IL 60607

Hollywood Look Alikes
3202 North 1300 East Road
Chebanes, IL 60922

Mickie Paskal/Rachel Tenner
Tenner-Paskal-Rudnicke
20 West Hubbard #2E
Chicago, IL 60610

E's Casting
440 West Barry
Chicago, IL 60657

Star Casting Network
770 North Halsted
Chicago, IL 60607

June Pyskacek
All-City Casting
PO Box 577640
Chicago, IL 60657

Jumpstart Casting & Production
4128 North Sheridan
Chicago, IL 60613

Trap Door Casting
1644 W. Courtland
Chicago, IL 60622

Shirley Hamilton
Shirley Hamilton, Inc
333 East Ontario #302B
Chicago, IL 60611

Non-Union Paige/Brett Ehlman
McBlaine & Associates
805 W. Touhy Avenue
Park Ridge, IL 60068

David O'Connor Casting
1017 West Washington #2A
Chicago, IL 60607

Bonnie Murray/David Lewis
Voice Casting, Inc.
400 S. Green Street #502
Chicago, IL 60707

Hix Casting
760 Mather Lane
Batavia, IL 60510

Jennifer Rudnicke
Tenner-Paskal-Rudnicke
20 West Hubbard #2E
Chicago, IL 60610

Voice Casting – Casting Dept.
Big House Casting & Audio
944 North Noble
Chicago, IL 60622

Beth Rabedeau
Rabedeau Casting
920 North Franklin #205
Chicago, IL 60610

John King
John King Productions
244 Madison Avenue #393
New York, NY 10016

Linda Jack
Linda Jack Talent
230 East Ohio #200
Chicago, IL 60611

Native American Talent
Neyanta
PO Box 170325
Chicago, IL 60617

Claudine Cansu
Royal Model & Talent Mgmt
1051 Perimeter Drive #530
Schaumburg, IL 60173

Maureen Brockman
Stewart Talent Management
58 West Huron
Chicago, IL 60610

Vo-Sharon C. Wottrich
Voices Unlimited Inc
541 North Fairbanks #27th Fl
Chicago, IL 60611

Ford Talent Group
1017 W Washington, Suite 2C
Chicago, IL 60640

Elizabeth Geddes
Geddes Agency
1633 North Halsted St #300
Chicago, IL 60614

Impact Talent Assoc Inc
333 West Wacker Drive
Chicago, IL 60606

Japanese Communications
333 North Michigan Ave. #411
Chicago, IL 60601

Lori Lins
Lori Lin's LTD Talent Mgmt
7611 East Holmes Avenue
Milwaukee, WI 53220

8 Talent & Casting
20 N Hubbard #2E
Chicago, IL 60690

Nouvelle Talent, Inc
PO Box 578100
Chicago, IL 60657

Myrna Salazar
Salazar & Navas, Inc #700
700 N Green Street #503
Chicago, IL 60622

Non-Union B. Boden/J. Wilson
Talent Group
4755 North Hermitage Avenue
Chicago, IL 60640

Arlene Wilson
Arlene Wilson Talent
430 West Erie #210
Chicago, IL 60610

Jacquelyn Conard
JSA-RAH Casting
700 North Green, Suite C-20
Chicago, IL 60622

Jennifer L. Berg
Jennifer's Talent Unlimited Inc
740 North Plankinton #300
Milwaukee, WI 53203

Joan Brown
ETA Inc
7558 South Chicago Avenue
Chicago, IL 60619

Voice Casting-Debby Kotzen
Naked Voices
865 North Sangamon, Suite 415
Chicago, IL 60622

John Robert Powers Modeling
8430 W Bryn Maur Ave. Ste 100
Chicago, IL 60631

Non Union-Karen Stavins
Karen Stavins Enterprises
Concourse Level
303 East Wacker Drive
Chicago, IL 60601

Talent Undiscovered
5647 West Irving PK
Chicago, IL 60634

Ford Talent Group Inc
641 West Lake, Suite 402
Chicago, IL 60661

Ford Talent Group
4755 North Hermitage Avenue
Chicago, IL 60640

Act One
640 North LaSalle Dr #535
Chicago, Il 60610

BMG Model & Talent
456 North May Street
Chicago, IL 60622

John Casablancas
Modeling & Career Center
200 N Patrick Blvd, Suite 700
Brookfield, WI 53045

New Talent
Encore Talent Agency, Inc
700 N. Sacramento Blvd #221
Chicago, IL 60612

Susan Sherman
Ambassador Talent Agents, Inc
333 N. Michigan Ave. #314
Chicago, IL 60601

Diane Rowley
Baker & Rowley Talent Agency
1200 West 35th Street
Chicago, IL 60609

Marie A. Boyd
Robert Schoeder
Aria Model & Talent Mgmt
333 North Michigan Ave #910
Chicago, IL 60601

FLORIDA AGENTS & CASTING DIRECTORS

Meris L. Zittman
International Artists Group, Inc
2121 N. Bayshore Dr. #2E
Miami, FL 33137

Robert G. Kreusler/Autumn McFadden
Wilhelmina-Miami
930 Washington 4th Floor
Miami Beach, FL 33139

Paul Levine
Famous Faces Company Inc
3780 SW 30th Avenue
Ft. Lauderdale, FL 33312

Beverly Sage
Runways, The Talent Group
1688 Meridian Avenue #500
Miami Beach, FL 33139

Tammy Green
The Green Agency Inc
1329 Alton Road
Miami Beach, FL

Cynthia Sheffield
The Sheffield, Inc
800 West Avenue Suite C1
Miami, Fl 33139

Elan Model talent Mgmt & Casting
899 South Plymouth Court
Chicago, IL 60605

Brooke Tonneman
Big Mouth Talent Agency
101 West Grand, Suite 200
Chicago, IL 60610

Dupont Plaza-Josh Gillet, Arthur
Arthur Inc.
300 Biscayne Blvd. Way #723
Miami, FL 33572

Enid Howell
Marian Polan Talent Agency
10 NE 11th Avenue
Ft Lauderdale, FL 33301

Kids/Young Adults/Rosie Delarosa
Age Kids & Teens Modeling Mgmt.
1210 Washington Avenue #250
Miami Beach, FL 33139

Irene Marie
Irene Marie Inc
728 Ocean Drive
Miami Beach, FL 33139

Debra Cozza
World Of Kids Inc
1460 Ocean Drive #205
Miami Beach, FL 33139

Roxanne McMillan
Roxanne McMillan Talent
12100 NE 16th Avenue #106
Ft Lauderdale, FL 33161

Alliance Talent Group Inc
1940 Harrison Street Suite 300
Hollywood, Fl 33020

Nouvelle Talent, Inc
PO Box 578100
Chicago, IL 60657

Elite Model Mgmt
58 West Huron
Chicago, IL 60610

Cathy Pearson Tully
Coconut Grove Talent Agency
3525 Vista Court
Coconut Grove, FL 33133

Cindy Freed
Stellar Talent Agency
407 Lincoln Road #2K
Miami Beach, FL 33139

Danielle Bienkievitz/Jessica Grace Black
Alliance Talent Group, Inc
10200 State Road 84 #211
Davie, FL 33325

Sharon Martin
Martin & Donalds Talent
2131 Hollywood Blvd. #308
Hollywood, FL 33020

Anita Spiegel
Boca Talent & Model Agency
829 SE 9th Street
Deerfield Beach, FL 33441

Christine Donalds
Martin & Donalds Talent
2131 Hollywood Blvd. #308
Hollywood, FL 33020

CENTRAL FLORIDA AGENTS

Susan Scher
Alexa Model & Talent Inc
4100 W Kennedy Blvd. #228
Tampa, FL 33609

Linda Koldenhoven
The Christensen Group
4395 St. Johns Pkwy
Sanford, FL 32771

Sandi Bell
The Sandi Bell Talent Agency
2582 S. Maguire Road #171
Ocoee, FL 34761

Traci Danielle
Brevard Talent Group
301 E Pine Street Suite 175
Orlando, Fl 32801

Louise Kahle
Louise's People Model & Talent
863 13fth Avenue N.
St. Petersburg, FL 33701

Marya Pemberton
Central Florida Talent, Inc
2601 Wells Avenue #181
Fern Park, FL 32730

Dawn Juergensen-Garrido
Outlook Orlando, Inc
851 State Road 436 #1057
Altamonte Springs, FL 32714

Kathy Laughlin
Independent Casting
Bldg. 200 #250
8313 W Hillsborough Avenue
Tampa, FL 33615

Kimberly Mullen
Kimberly/Mark Mullen Casting
4390 35th Street
Orlando, FL 32811

Steve Benz
Benz Model & Talent
The Firehouse
1313 8th Ave 3rd Floor
Tampa, FL 33605

Evelyn Stewart
Evelyn Stewart Model/Talent Agency
911 Samy Drive
Tampa, FL 33613

Diana Arthur
Diana Arthur Inc
6542 US Hwy 41 N. #205A
Apollo Beach, FL 33572

Historic District-Meghan Drewett
The Diamond Agency, Inc
585 E State Road #434
Longwood, FL 32750

Kirsten Dalrymple
Azuree Talent Agency
140 N Orlando Avenue Suite 120
Winter Park, FL 32789

Edna Byers
Dimensions III Talent Agency
5205 S. Orange Avenue #209
Orlando, FL 32809

FLORIDA CASTING DIRECTORS

Ellen Jacoby
Ellen Jacoby International Casting
150 South East 2nd Ave Suite 1007
Miami, FL 33131

Pati Robinson
The Reality Group, Inc
1185 Bay Shore Drive
Englewood, FL 34223

Nicole Reinholt/Gregory Brown
BMG Models
2610 Corrine Drive
Orlando, FL 32803

Vicki L Foley
In Any Event
140 S. Atlantic Avenue 5th FL
Ormond Beach, FL 32176

Carol Berg
Berg Talent & Model Agency
15908 Eagle river Way
Tampa, FL 33624

Kim Cinque
Florida Stars Model/Talent
1020 NW 23rd Ave.
Gainesville, FL 32609

Greta Hanley
BI Coastal Talent
398 W Amelia St.
Orlando, FL 32801

Melanie Hurt
The Hurt Agency, Inc
400 N. New York Avenue #207
Winter Park, FL 32789

Dee Miller
Casting Directors Inc
7300 NE 4th Ct.
N. Miami, FL 33138

Brad Davis
Marquee Entertainment Casting
9705 NE 2nd Avenue
Miami, Fl 33138

RJ Istvan
RJ's Casting
656 Bellhurst Drive
Orlando, FL 32835

Barbara Diprima
Diprima Casting Inc
Hotel Biscaya Top Floor
4000 Alton Road
Miami Beach, FL 33139

Robert P. Kahn
For-Casting
PO Box 593763
Orlando, FL 32859

Lillian Gordon
Manhattan South Casting
1012 N Mills Avenue
Orlando, FL 32803

Beverly McDermott
Beverly McDermott Casting
923 N. Golf Drive
Hollywood, FL 33021

Thomas Lech
Lechnology Casting
2815 Auburn Avenue W.
Tampa, FL 33614

Greenwich Studios-T R Durphy
T R Durphy Casting
12100 NE 16th Avenue #110
N. Miami, FL 33161

Ed Arenas
Unique Casting & Associates
1613 Alton Road 2nd Floor
Miami Beach, FL 33139

Ellen Jacoby Casting
420 Uncoln Road Suite 210
Miami Beach 33139

Kim Houston
Houston Kasting
PO Box 53-1101
Miami, FL 33153

Ellen Jacoby
Ellen Jacoby Casting Intl Ltd
300 Biscayne Blvd. Way #1150
Orlando, FL 33131

John King
John King Productions
244 Madison Avenue #393
New York, NY 10016

George Ortiz
Miami Talento Casting Inc
335 Almeria Ave
Coral Gables, FL 33134

Melvin Johnson
Melvin Johnson & Associates
Bldg 22 Room 235
1000 Universal Studio Plaza
Orlando, Fl 32819

Lori Wyman
Lori Wyman Casting
16499 NE 19th Avenue Suite 203
N. Miami, FL 33162

LA SOAPS AGENTS & CASTING DIRECTORS

Rick Jacobs
Lifetime TV
2049 Century PK E. #840
Los Angeles, CA 90067

Shannon Bradley (B) CBS-TV City
The bold And The Beautiful
7800 Beverly Blvd #3371
Los Angeles, CA 90036

Mark Teschner-General Hospital
ABC-TV
4151 Prospect Ave. Sta. 54
Los Angeles, CA 90027

Principal Work-Marnie Saitta
CBS-The Young & The Restless
7800 Beverly Blvd. #3305
Los Angeles, CA 90036

Meg Liberman
Liberman/Patton Casting
6464 Sunset Blvd. #707
Los Angeles, CA 90028

Tammy Billik
Tammy Billik Casting
14044 Ventura Blvd. #309
Sherman Oaks, CA 91604

Senior VP Talent/Casting
Mary V Buck
Warner Brothers Bldg. #140
300 Television Plaza. 1st FL.
Burbank, CA 91505

Andrea Cohen
Andrea Cohen Casting
606 N. Larchmont Blvd. #8
Los Angeles, CA 90004

Melinda Gartzman
Melinda Gartzman Casting
11271 Ventura Blvd #248
Studio City, CA 91604

Vicki Huff
Vicki Huff Casting
5700 Wilshire Blvd #500 N
Los Angeles, CA 90036

Fran Bascom-Days Of Our Lives
Columbia Picture (TV)
3400 Riverside Drive #765
Burbank, CA 91505

Gwen Hillier-General Hospital
ABC-TV
4151 Prospect Ave. Sta. 54
Los Angeles, CA 90027

Don Philip Smith
NBC-Casting/Passions
4024 Radford Ave. #280
Studio City, CA 91604

Reuben Cannon
Reuben Cannon & Associates
5225 Wilshire Blvd #526
Los Angeles, CA 90036

Christy Dooley/CBS-TV City
The Bold And The Beautiful
7800 Beverly Blvd #3371
Los Angeles, CA 90036

Jackie Briskey
NBC-Casting/Passions
4024 Radford Ave #280
Studio City, CA 91604

Escape Artists-Astaire Bldg
Denise Chamian Casting
10202 W Washington Blvd #3010
Culver City, CA 90323

Lisa Miller Katz
Warner Brothers
4000 Warner Blvd RM 106B
Burbank, CA 91522

Junie Lowry-Johnson
Bochco Bldg
20th Century Fox Studios
10201 At Avenue of the Stars #232
Century City, CA 90035

Fern Orenstein
CBS-VP Talent & Casting
7800 Beverly Blvd #284
Los Angeles, CA 90036

Nikki Valko
Valko Miller Casting
Manhattan Beach Studios
1600 Rosecrans Ave Bldg 4A 2nd Fl
Manhattan Beach, CA 90266

Roin Lippin
Robin Lippin Casting
606 North Larchmont Blvd 4B
Los Angeles, CA 90004

Man Dutton
Bldg 7 #110
Manhattan Beach Studios
1600 Rosecrans Ave.
Manhattan Beach, CA 90266

Linda Poindexter
Days Of Our Lives
Columbia Pictures (TV)
3400 Riverside Drive #765
Burbank, CA 91505

Deborah Barylski
Bldg 72nd Fl
CBS Studio Center
4024 Radford Ave
Studio City, CA 91604

Mali Finn
Mali Finn Casting
303 N Sweetzer Avenue
Los Angeles, CA 90048

Tracy Lilienfield (Sound Shop)
CBS Studio Center
4024 Radford Ave 2nd Fl
Studio city, CA 91604

Francine Maisler
Tri-Star Bldg
Sony Pictures
10202 W Washington Blvd #207
Culver City, CA 90323

Clair Sinnett
Clair Sinnett Casting
531 Main St. #1135
El Segundo, CA 90232

Geri Windsor
Geri Windsor Casting
606 N Larchmont Blvd #4B
Los Angeles, CA 90004

Susan Bluestein
Bldg 35 4th FL
Sunset/Gower Studios
1438 N Gower St. #453
Los Angeles, CA 90028

Jeff Greenberg
Marx Brothers Bldg
Paramount Studios
5555 Melrose Ave #102
Los Angeles, CA 90038

Gail Camacho-Background
CBS-The Young & The Restless
7800 Beverly Blvd #3305
Los Angeles, CA 90036

Deedee Bradley
Deedee Bradley Casting
6767 Forest Lawn Drive #100
Los Angeles, CA 90068

Sheila Guthrie
Bludohen Bldg
Paramount Studios
5555 Melrose Ave #102
Los Angeles, CA 90038

John Levey
VP Casting RM 106
Warner Bros. TV Center
4000 Warner Blvd Bldg 17
Burbank, CA 91505

Julie Mossberg
Mossberg/Anthony Casting
4024 Radford Ave. TR. 800
Studio City, CA 91604

Sally Stiner
Culver Studios
Stiner/Block Casting
9336 W Washington Blvd St 11 #1
Culver City, CA 90232

Bill Dance/818-754-6634
Bill Dance Casting
4605 Lankershim Blvd #401
N. Hollywood, CA 91602

Michael Hothorn, Dir. Casting #448
20th Century Fox Studios Bldg 89
10201 At Avenue of the Stars
Century City, CA 90035

Tony Sepulveda
Bldg 140 1st Fl
Warner Bros TV
300 TV Plaza
Burbank, CA 91505

Dava Waite
Bldg 7 2nd Fl
CBS/Dava Waite Casting
4024 Radford Ave
Studio City, CA 91604

Debi Manwiller
Pagano/Manwiller Casting
6250 Canoga Avenue
Woodland Hills, CA 91367

Marcia Shulman
SR. VP Talent/Casting
20th Century Fox TV
10201 W Pico Blvd #4503 Bldg 100
Los Angeles, CA 90035

April Webster
April Webster Casting
800 S Main St #309
Burbank, CA 91506

Pamela Wagner
Agency For Performing Arts
9200 Sussent Blvd #900
Los Angeles, CA 90069

Bonnie Black
Bonnie Black Talent Agency
12034 Riverside Drive #103
Valley Village, CA 91607

Craig Wyckoff
Epstein-Wyckoff-Corsa-Ross & Assoc.
280 S Beverly Drive #400
Beverly Hills, CA 90212

Megan McConnell
Gilmore/McConnell Casting
Manhattan Beach Studios
1600 Rosecrans Ave Bldg $B 1st Fl
Manhattan Beach, CA 90266

Ronnie Yeskel
Ronnie Yeskel Casting
3231 Ocean Park Blvd #212
Santa Monica, CA 90405

Kathleen Letterie
Warner Bros TV Center RM161
3701 W Oak St Bldg 34R
Burbank, CA 91505

Lisa Mionie
Lisa Mionie Casting
606 N Larchmont Blvd #4B
Los Angeles, CA 90004

Dawn Stenberg
SR VP Talent/Casting RM 207
9336 W Washington Blvd Bldg C
Culver City, CA 90232

Bonnie Zane
Zane/Pillsbury Casting
585 N Larchmont Blvd
Los Angeles, CA 90004

Robert Malcolm
The Artists Group #2490
10100 Santa Monica Blvd
Los Angeles, CA 90067

Kay Billings
Chateau/Billings Talent
5657 Wilshire Blvd #200
Los Angeles, CA 90036

Dale Garrick
Dale Garric Intl Talent
1017 N LaCienega Blvd #109
W. Hollywood, CA 90069

Robi Reed-Humes
Robi Reed & Associates
6605 Hollywood Blvd #100
Los Angeles, CA 90028

Joy Todd
MGM Plaza
Todd's Campobasso Casting
2425 Colorado Ave Bldg B #208
Santa Monica, CA 90404

Molly Lopata
Molly Lopata Casting
13731 Ventura Blvd #A
Sherman Oaks, CA 91423

Vicki Rosenberg
David Lean Bldg
Sony Studios
10202 W. Washington Blvd #100
Culver City, CA 90232

Robert Ulrich
Ulrich/Dawson/Kritzer
4705 Laurel Canyon Blvd Suite 301
Valley Village, CA 90401

LOS ANGELES AGENTS

J. Michael Bloom
Marc Bass Agency
415 N Crescent Drive #320
Los Angeles, CA 90210

Bert Charles
The Charles Talent Agency
11950 Ventura Blvd #3
Studio City, CA 91604

Gar Lester
JLA Talent
9151 Sunset Blvd
W. Hollywood, CA 90069

Joe Rice
Abrams Artists Agency
9200 Sunset Blvd 11th Fl
Los Angeles, CA 90069

David Shaul
Bauman, Redanty & Shaul
5;757 Wilshire Blvd #473
Los Angeles, CA 90036

Elaine Ellison
The Culbertson Group
8430 Santa Monica Blvd #210
Los Angeles, CA 90069

Gregg Meyer
Acme Talent
4727 Wilshire Blvd #333
Los Angeles, CA 90010

Jennifer Jones
Brand Talent & Model Agency
1520 Brookhollow Drive #39
Santa Ana, CA 92705

Linda Jenkins
Cunningham, Escott & Dipene
106 Santa Monica Blvd #130
Los Angeles, CA 90025

William Kerwin
William Kerwin Agency
1605 N. Cahuenga Blvd
Hollywood, CA 90028

IRV Schechter
IRV Schechter Company
9300 Wilshire Blvd #400
Beverly Hills, CA 90212

Diane Davis
Twentieth Century Artists
15760 Ventura Blvd #700
Encino, CA 91436

Holly Davis-Carter
Agency West Ent.
5750 Wilshire Blvd #640 N.
Los Angeles, CA 90048

Tim angle
Don Buchwald & Associates
6500 Wilshire Blvd #2200
Los Angeles, CA 90048

Tina Kiratsoulis
Flick East-West Talents
9057 Nemo St. #A
W. Hollywood, CA 90069

Pamela Porter
AKA Talent Agency
6310 San Vincente blvd #200
Los Angeles, CA 90048

Ray Cavaleri
Cavaleri & Associates
178 S Victory Blvd #205
Burbank, CA 91502

Cris Dennis
Film Artists Associates
4717 van Nuys Blvd #215
Sherman Oaks, CA 91403

Raphel Berko
Media Artist Group
6404 Wilshire Blvd #950
Los Angeles, CA 90048

Pam Sparks
Suton, Barth & Cennari
145 S Fairfax Avenue #310
Los Angeles, CA 90036

Shirley Wilson
Shirley Wilson Agency
5410 Wilshire Blvd #806
Los Angeles, CA 90036

Kim Dorr
Bobby Ball Talent Agency
4605 Lankershim Blvd #721
Universal City, CA 91602

Amber Raitz
Coast To Coast Talent Group Inc
3350 Barham Blvd
Los Angeles, CA 90068

Michael Linden Greene
Greene & Associates
7080 Hollywood Blvd Suite 1017
Hollywood, CA 90028

Lyn Baldwin
Baldwin Talent Inc
8055 W Manchester #550
Los Angeles, CA 90048

Randi Rubenstein
Commercials Unlimited Inc
8383 Wilshire Blvd #850
Beverly Hills, CA 90211

Teresa Valente
Beverly Hecht Agency
12001 Ventura Pl #320
Studio City, CA 91604

Marjorie Sperling
Pacific West
12500 Riverside Dr. #202
N. Hollywood, CA 91607

Scott Stander
Scott Stander :& Associates
13701 Riverside Drive #201
Sherman Oaks, CA 91423

Central Casting
220 S. Flower St
Burbank, CA 91502

Dick Lovell
Lovell & Associates
7095 Hollywood Blvd #1006
Los Angeles, CA 90028

William Calder
Special Artists Agency
345 N Maple Drive #302
Beverly Hills, CA 90210

Susan Jones
US Talent
485 S Robertson Blvd #7
Beverly Hills, DA 90211

Nick Terzian
NTA TALENT AGENCY
8899 BEVERLY BLVD #612
LOS ANGELES, CA 90048

Helen Garrett
The Savage Agency
6212 Banner Avenue
Los Angeles, CA 90038

Mary Jo Slater
Slater/Brooksbank
800 S. Main St. 2nd FL
Burbank, CA 91521

Suzanne Crosby
Image Makers
119 Braintreet Street #109
Allston, MA 02134

T. Denard/S. Garrett
McDonald Richards, Inc
5 River Road #317
Wilton, CT 06897

Esther Johnston
Johnston Agency
50 Washington St.
S. Norwalk, CT 06854

Susan Nathe
Susan Nathe & Associates
8281 Melrose Avenue #200
Los Angeles, CA 90046

Samantha Botana
The Samantha Group
300 S Raymond Avenue #11
Pasadena, CA 91105

Toni Susskind
J S Represents
936 ½ North La Jolla Avenue
Los Angeles, CA 90046

Jack Scagnetti
Jack Scagnetti Talent Agency
5118 Vineland Avenue #102
N. Hollywood, CA 91601

Elton Bolden
Tilmar Talent Agency
4929 Wilshire Blvd #1020
Los Angeles, CA 90010

BOSTON AGENTS

Maggie Trichon
Maggie, Inc.
35 Newbury St.
Boston, MA 02116

Model Club, Inc
269 S. Main St.
Providence, RI 02903

Ed Sliney
Model Club, Inc
329 Columbus Ave
Boston, MA 02116

Alosa Adler
Pardigm 25th Fl
10100 Santa Monica Blvd
Century City, CA 90067

Tracy Mapes
Arlene Thornton & Associates
12701 Ventura Blvd #201
Studio City, CA 91604

Brian S McCabe
McCabe/Justice
8285 Sunset Blvd #1
Los Angeles, CA 90046

Tim Stone
Stone-Manners Agency
8436 West 3rd Street Suite 740
Los Angeles, CA 90048

Bob Waters
Bob Waters Agency
9301 Wilshire Blvd #300
Los Angeles, CA 90210

Lynda St. James
The Cambo Agency
49 River St. #1
Waltham, M 02453

Donahue Models
14 Rome Ave.
N. Providence, RI 02908

Ford Model Mgmt, Inc
297 Newbury St.
Boston, MA 02115

BOSTON CASTING DIRECTORS

Peter Bezemes
Peter Bezemes Casting
28 State Street, 11th Floor
Boston, MA 02109

R.I. Casting Service
807 Broad St.
Providence, RI 02907

Kevin Fennessy
Kevin Fennessy Casting
165 Amory Street
Cambridge, MA 02139

Maura Tighe
Tighe & Doyle Casting
142 Berkley St. 4th FL
Boston, MA 02116

New England Model Group
55 South Commercial St
Larger Place
Manchester, NH 03101

Julie Arvedon
NE/BTS
501 Boylston St.
PO Box 9021
Boston, MA 02116

Rena Maliszewski
Hearst-Argyle TVG Prod
5 TV Place
Needham, ma 02194

Casting Dept
Chedd-Angler
70 Coolidge Hill Road
Watertown, MA 02472

Casting Dept
Picture Park
811 Boylston Street
Boston, MA 02116

Tom Elliott
Tom Elliott Productions
PO Box 441
Waltham, MA 02154

Susan Willett
Susan Willett Casting
65 King Street
Swampscott, MA 01907

Anne Mulhall
LDI Casting
PO Box 7105
Warwick, RI 02886

Carolyn Pickman
C P Casting
537 Tremont St.
Boston, MA 02116

Casting Dept
Studio 1
P O Box 580
Rehoboth, MA 02769

Steve Stone
Creative Video
93 West Street
Medfield, MA 02180

Audio Tapes – Casting Dept
RRB Production Company
4 Charlotte Road
Newton, MA 02459

Jennifer Wry
Commonwealth Media Inc
223 Commonwealth Avenue
Boston, MA 02116

Casting Dept
Replica Video Productions
143 Pleasant Street #3A
Cambridge, MA 02139

John King
John King Productions
244 Madison Ave #393
New York, NY 10016

Angela Peri
Boston Casting
129 Braintree St. #107
Boston, MA 02134

VO Tapes Only Casting Dept
Soundtrack/Cav Corp
162 Columbus Ave
Boston, MA 02116

Lori Franklin
Lori Franklin Casting
Astor Station
P O Box 475
Boston, MA 02123

BOSTON AUDIO/FILM/VIDEO

Patrick Coppola
Coppola Ent./5 Angeles Production
26 Lincoln Street
Stoneham, MA 02194

Jared Katsiane
Straight Cut Productions
39 Clarendon Street
Boston, MA 02134

Jane Picor
WGBH Production
125 Western Avenue
Boston, MA 02134

Casting Dept
Scout Productions
119 Braintree Street
Allston, MA 02116

Jill Petrie
Cramer Productions
425 University Avenue
Norwood, MA 02062

Casting Dept
Potter Ruiz Productions
37 Robin Hood Road
Arlington, MA 02174

Jim Amadeo
Clarke Goward Advertising
535 Boyleston Street #8
Boston, MA 02116

Judy Blye-Wilson
ABC-All My Children
320 W 66th Street
New York, NY 10023

Casting Dept. (B)
Cecelia Pleva
NBC-Conan O'Brien Show
30 Rockefeller PL #90W
New York, NY 10112

Hack
Mark Saks
CBS-TV Primetime Casting
51 W 52nd Sta. 20th FL
New York, NY 10019

Julie Madison
ABC-One Life To Live
157 Columbus Avenue 2nd FL
New York, NY 10023

Vinnie Postestivo
MTV/MTV2
1515 Broadway 25th FL
New York, NY 10036

Meg Simon
Fox-TV
1211 Sixth Avenue 3rd FL
New York, NY 10036

Gregory Sullivan
Cooper Productions, Inc
100 Messina Drive #E
Braintree, MA 02184

BOSTON ADVERTISING AGENCIES

Steve Ruggieri
Donnelly & Duncan
865 Waterman Avenue
E. Providence, RI 02914

Robert Lambert
ABC-All My Children
320 W 66th Street
New York, NY 10023

JC Studios
Mary Clay Boland
CBS-As The World Turns
1268 E. 14th Street
Brooklyn, NY 11230

JC Studios Assistant
Kate Martineau
CBS-As The World Turns
1268 E. 14th Street
Brooklyn, NY 11230

Rosalie Joseph
Primetime Casting
ABC-TV
157 Columbus Avenue 2nd FL
New York, NY 10023

Casting Dept. (B)
Josh Payne
NBC-Saturday Night Live
30 Rockefeller PL
New York, NY 10112

Naomi Frisch
Comedy Central
1775 Broadway 6th FL
New York, NY 10019

Peter & Bobby Gilmore
Gilmore Associates
123 Lewis Wharf
Boston, MA 02109

Leo Rice
Christopher Thomas Assoc
240 Newbury Street
Boston, MA 02116

NEW YORK SOAPS/SERIES/ PRIMETIME

Sheryl Fisher
ABC-One Life To Live
157 Columbus Avenue 2nd FL
New York, NY 10023

Courtney Mullin
MTV/TRL
1515 Broadway 23rd FL
New York, NY 10036

JC Studios
Lamont Craig
CBS-As The World Turns
1268 E. 14th Street
Brooklyn, NY 11230

Anthony Volastro
ABC-All My Children
320 W 66th Street
New York, NY 10023

ED (P)
James Calleri
Playwrights Horizons
416 West 42nd Street
New York, NY 10036

Amy Herzig
Primetime Casting
CBS-TV
51 W 52nd Sta. 20th FL
New York, NY 10019

Janet Murphy Butler
ABC-TV
157 Columbus Avenue 2nd FL
New York, NY 10023

#1265E
Steven O'Neill
NBC-TV
30 Rockefeller PL
New York, NY 10112

Whoopi (B)
Byron Crystal
Byron Crystal Casting
41 Union Square W #316
New York, NY 10003

Rob Decina
CBS-The Guiding Light
222 E 44th Street 4th FL
New York, NY 10017

Andra Reeve
CBS-TV
51 W 52nd Sta. 14th FL
New York, NY 10019

Hack (B)
Diane Heery
Heery Casting
230 North 2nd St. #1A
Philadelphia, PA 19106

Third Watch, Law & Order SVU (B)
Casting Dept
Grant Wilfley Casting
60 Madison Ave. #1027
New York, NY 10010

Criminal Intent (P)
G Keller
Lynn Kressel Casting Pier 62 #30
Hudson River W 23rd Street #308
New York, NY 10011

Sex & The City (P) Hiatus
Meredith J Marciano
Amerifilm Casting Silver Cup
42-22 22nd St. #200
Long Island City, NY 11101

Marci Phillips
Primetime Casting
ABC-TV
157 Columbus Avenue 2nd FL
New York, NY 10023

Casting Dept. (P)
M Klein/A Cohen
NBC-Saturday Night Live
30 Rockefeller PL
New York, NY 10112

Law & Order (P)
Suzanne Ryan
Lynn Kressel Casting Pier 62 #30
Hudson River W 23rd Street
New York, NY 10011

Alison Rinzel-Lucid
Daytime Casting
CBS-TV
51 W 52nd Sta. 14th FL
New York, NY 10019

Melanie Haseltine
CBS-The Guiding Light
222 E 44th Street 4th FL
New York, NY 10017

Ed Sullivan Theatre
Jill Lieberman
Late Show/David Letterman
1697 Broadway Studio 50
New York, NY 10019

The Sopranos (P)
Georgianne Walken
Walker/ Jaffe
729 Seventh Avenue #14A
New York, NY 10019

Nevitt Greene
Americas Most Wanted
2 Bethesda Metro Center #800
Bethesda, MD 20814

Sex & The City (P) Hiatus
Jennifer McNamara
Silver Cup Studios
42-22 22nd St. #200
Long Island City, NY 11101

Tori Visgilio
ABC-One Life To Live
157 Columbus Avenue 2nd FL
New York, NY 10023

Wire (P)
Alexa Fogel
Beech Hill Films
450 West 15th Street #602
New York, NY 10011

Whoopi (P)
Bernard Telsey
Bernard Telsey Casting
145 W 28th Street #12F
New York, NY 10001

Law & Order (P)
Jonathan Strauss
Lynn Kressel Casting Pier 62 #30
Hudson River W 23rd Street
New York, NY 10011

Lara Tal
Primetime Casting
CBS-TV
51 W 52nd Sta. 14th FL
New York, NY 10019

Law & Order (B)
Casting Dept.
Sylvia Fay Casting
71 Park Avenue
New York, NY 10016

Hope & Faith (B)
Karen Etcoff
Kee Casting
424 Park Avenue So. #128
New York, NY 10016

Third Watch (P)
B Bowling/ K Miscia Casting
159 W 25th St #1017
New York, NY 10001

Hope & Faith (P)
Jennifer McNamara
Silver Cup Studios
42-22 22nd St. #200
Long Island City, NY 11101

Background
Ricardo Bertoni
Navarro/Bertoni Casting Co.
101 W 31st St #912
New York, NY 10001

Juliet Taylor
Juliet Taylor Casting
140 W. 57th Street #4B
New York, NY 10019

Stephanie Klapper
Stephanie Klapper Casting
122 W 26th Street #1104
New York, NY 10001

Mike Lemon
Mike Lemon Casting
413 N 7th Street #602
Philadelphia, PA 19123

Loren Lightenstein
Loren Lightenstein Casting
163 River Road
Preston, CT 06365

3 of US Studios
Joan Lynn
Joan Lynn Casting
39 W 19th Street
New York, NY 10011

David Caparelliotis
Nancy Piccione
Manhattan Theatre Club
311 W 43rd Street 8th FL
New York, NY 10036

Alan Filderman
Alan Filderman Casting
333 W 39th Street #601A
New York, NY 10018

Janet Foster
Janet Foster Casting
3212 Cambridge Avenue
New York, NY 10463

Marie E Nelson
Ellyn Long Marshall
Orpheus Group, Inc. (Casting)
165 W 46th St #405
New York, NY 10001

Ricordo Codero
Global Network Pictures
244 Madison Avenue 2nd Fl #A215
New York, NY 10001

Cinema 7
Robyn Knoll
Robyn Knoll Casting
154 W 57th Street #112
New York, NY 10019

Indigent
Ellen Lewis Casting
105 Greene Street Fourth FL
New York, NY 10012

Tom Alberg/Vince Liebhart
Liebert/Alberg Casting
1710 First Avenue #122
New York, NY 10128

Amanda Mackey Johnson
Sig DeMiguel
Mackey/Sandrich Casting
145 West 28th Street Suite 12F
New York, NY 10001

Joe McConnell
Joe McConnell Casting
145 W 28th Street Suite 12F
New York, NY 10001

Leonard Finger
Leonard Finger Casting
1501 Broadway #1511
New York, NY 10036

Overland Ent.
Kelly Foster
Kelly Foster Casting
257 W 52nd Street 5th FL
New York, NY 10019

NEW YORK CASTING DIRECTORS

Avy Kaufman
Avy Kaufman Casting
180 Varick Street 16th FL
New York, NY 10014

Jodi Kipperman
Kipperman Casting, Inc
12 W 37th Street 3rd FL
New York, NY 10018

Ericka Kurzman
Andrea Kurzman Casting
122 E 37th Street 1st FL
New York, NY 10016

Liz Lewis
Liz Lewis Casting
129 W 20th Street
New York, NY 10011

Jennifer Low Sauer
Jennifer Low Sauer Casting
c/o Keen Company
520 8th Avenue #328
New York, NY 10018

Joel Manaloto
Joel Manaloto Casting
1480 York Avenue 4th FL
New York, NY 10021

Indigent
Susie Farris
Susie Farris Casting
135 W 26th Street 5th FL
New York, NY 10001

Denise Fitzgerald
Denise Fitzgerald Casting
111 East 12th Street 2nd Fl
New York, NY 10003

Doreen Frumkin
Doreen Frumkin Casting
140 W 22nd Street
New York, NY 10011

Andrea Shane
Brette A. Goldstein
Shane/Goldstein Casting
311 W 43rd Street #602
New York, NY 10036

Lori Eastside
NY City Casting
212 Bowery 3rd FL
New York, NY 10012

Helen Taylor
Helen Taylor Casting
140 W 58th St #8A
New York, NY 10019

Carole M. Duckworth
Dan Duckworth
AAA Voicecasting
123 W 18th St 7th FL
New York, NY 10011

Mike Pepino
Mike Pepino Casting
PO Box 57
Stanton, NJ 08885

Jerry Beaver
Jerry Beaver & Associates
215 Park Avenue Sout
New York, NY 10003

David Clemmons
David Clemmons Casting
165 W 46th Street Suite 1100
New York, NY 10036

Rich Cole
Rich Cole Casting
648 Broadway #912
New York, NY 10012

Jessica Gilburne
Gilburne/Urban Casting
80 Varick Street #6A
New York, NY 10013

Linda Godlove
Godlove Company Casting
151 W 25th Street 11th FL
New York, NY 10001

Sam/Susan Gish
Philadelphia Casting Company
114 Chestnut St 3rd FL
Philadelphia, PA 19106

Jennifer Rudolph
Jeff Mitchell
Mitchell/Rudolph Casting
440 Park Avenue S 11th FL
New York, NY 10016

Stephanie Holbrook
Doug Aibel
Vineyard Theatre
108 E 15th St
New York, NY 10003

P Kerrigan Dicerto
Juliet Taylor Casting
140 W 57th St #4B
New York, NY 10019

Breanna Benjamin
Breanna Benjamin Casting
406 W 31st Street
New York, NY 10001

Paul Fouquet
Elissa Myer
Elissa Myers Casting
333 W 52nd St
New York, NY 10019

Lisbeth Andresen/Jeff Barber
CTP Casting
207 W 25th St 6th Floor
New York, NY 10001

Maureen Fremont
Maureen Fremont Casting
150 W 80th Street #4C
New York, NY 10024

David Vaccari
Bernard Telsey
Bernard Telsey Casting
145 W 28th St #12F
New York, NY 10001

Carol Nadell
Selective Casting/Carol Nadell
Radio City Sta. POB 1538
New York, NY 10101

Casting Dept
Amhalise Casting
111 E 14th St. #251
New York, NY 10003

Domenic Andreoli
Domenic Andreoli Casting
545 8th Avenue #401
New York, NY 10018

Background
Ulysses Terrero
T&T Casting
88-30 162nd St
Jamaica, NY 11432

Rori Floyd
Uniworld Group
100 Avenue of the Americas
New York, NY 10013

Jackie Kelly
Complete Casting
350 W 50th St
New York, NY 10019

Sue Crystal
Sue Crystal Casting
251 W 87th St #26
New York, NY 10024

M Delmonte
Merry L Delmonte Casting
575 Madison Avenue 10th FL
New York, NY 10022

Jennifer Joy Arnett
Sirius Casting
29 John St. PMB 126
New York, NY 10038

Stacy Osnow
Hungry Man
Downstairs Casting
428 Broadway 5th FL
New York, NY 10013

Martine Dorfman
Martine Dorfman Casting
163 Third Avenue #176
New York, NY 10003

Lisbeth White/Lucy Baker
Donna De Seta Casting
525 Broadway 3rd FL
New York, NY 10012

Bonnie Finnegan
Bonnie Finnegan Casting
12 W 27th St 11th FL
New York, NY 10001

VO Only Adam Goodman
Voicehunter.com
10 Sunset Drive
Weston, CT 06883

Tara Rubin
Tara Jayne Rubin
311 W 43rd St 5th FL
New York, NY 10036

SAG Only
Frank Disalvo
Ferrell & Calvillo Commun. Inc
250 Park Avenue S 11th Fl
New York, NY 10001

Steve Schaeffer
Donna De Seta
Donna De Seta Casting
525 Broadway 3rd FL
New York, NY 10012

Sidra Smith
Sidra Smith Casting
60 W 129th St #7E
New York, NY 10027

Rita Powers/Sean Powers
Extras Casting By Booked
9 Desbrosses St #513
New York, NY 10013

Louis Digiaimo
Louis Digiaimo & Associates
214 Sullivan St #2C
New York, NY 10012

Heidi Griffith/Jordan Thaler
The Public Theatre
425 Lafayette St
New York, NY 10003

Katharina Eggmann
Bonnie Finnegan Casting
12 W 27th St 11th FL
New York, NY 10001

Meg Simon
Warner Brothers
1325 Sixth Avenue 32nd FL
New York, NY 10019

Mele Nagler/Jim Carnahan
Roundabout Theatre Company
W 39th St 12th FL
New York, NY 10001

No Mail Accepted
Joy Weber
Joy Weber Casting/One On One
126 W 23rd St
New York, NY 10011

Indigent
Marcia Debonis
Marcia Debonis Casting
135 W 26th St 12th FL
New York, NY 10001

Janice Wilde
Ilene Starger
Ilene Starger Casting
1220 Broadway #801A
New York, NY 10001

Robin Dunn
Ed Entertainment
PO Box 20791
New York, NY 10023

Pennie Dupont
Pennie Dupont Casting
36 Perry St
New York, NY 10014

Robin D Carus
Theatreworks/USA
151 W 26th St 7th FL
New York, NY 10001

Steve Garrin/Louise Moore
Videoactive Talent
1780 Broadway #804
New York, NY 10019

P Huffman/O Harris
Warner Brothers
1325 Sixth Avenue 28th FL
New York, NY 10019

VO Only
Theresa Kiernan
Howard Schwartz Recording
420 Lexington Avenue #1934
New York, NY 10170

Tara Rubin Casting
Chelsa Studios
151 W 36th Street Studio 12
New York, NY 10001

VO Only
Lesley Freedman
McHale Barone
30 Irving Pl 6th FL
New York, NY 10003

Cindy Tolan
Vineyard Theatre
108 East 15th Street
New York, NY 10003

Debbie Kowall
Scott Powers
Scott Powers Productions
135 West 29th Street Suite 404
New York, NY 10001

Brooke Thomas/Neil Myer
House Production Casting
450 W 15th St #202
New York, NY 10011

J T
J T Talent and Mgmt
440 W 41st St #9E
New York, NY 10036

Rori Floyd
Uniworld Group
100 Avenue of the Americas 15th FL
New York, NY 10013

VO Only
Maria L Pappalardo
BBDO Inc
1285 Sixth Avenue 6th FL
New York, NY 10019

Mark Preiser
Walter F Cameron Advertising
350 Motor Pkwy #410
Hauppauge, NY 11788

Tuffy Questell
T E C Casting
PO Box 859
Bronx, NY 10462

Doris Silk
Media Inc
11 Fulton Place
W Hartford, CT 06107

Dan Aron
No Soap Productions
936 Broadway 4th Floor
New York, NY 10010

Kerry Barden
Hopkins, Smith & Barden
451 Greenwich St 5th FL
New York, NY 10013

Suzanne Smith
Manhattan Ensemble Theater Inc
55 Mercer Street
New York, NY 10013

Ana Melara
Wing Latino Group
114 Fifth Avenue 11th FL
New York, NY 10003

NEW YORK AD AGENCIES

Josette Lata
Bartle Bogle Hegarty Inc
7 W 22nd St 9th FL
New York, NY 10010

Ziccardi Partners
Joe Fontana, Frierson, Mee
1700 Broadway 35th FL
New York, NY 10019

Final Frame
Ann Goulder
Ann Goulder Casting
150 W 22nd St 7th FL
New York, NY 10011

Jennifer Sukup
Pomann Sound
2 W 46th St Ph
New York, NY 10036

Arnold J Mungioli
Mungioli Theatricals Inc
207 W 25th St 6th FL
New York, NY 10001

Billy Hopkins
Manhattan Ensemble Theater Inc
55 Mercer Street
New York, NY 10013

Stuart Howard/Amy Schecter
Stuart Howard Associates Ltd
207 W 25th St 6th FL
New York, NY 10001

Lew Sherwood
Warren, Kremer & Paino
2 Park Avenue #1400
New York, NY 10016

Tracy Lentini
Avrett, Free, Ginsberg, Inc
800 Third Avenue 35th FL
New York, NY 10022

Frank O'Blac
Robert A Becker Inc
75 Ninth Avenue 2nd FL
New York, NY 10011

Maya Camin
Lipman, Richard, Greene
408 W 14th St
New York, NY 10014

David Schiavone
Grey Global Group
777 Third Avenue 10th FL
New York, NY 10017

Stephen De Angelis
Grey Global Group
777 Third Avenue 10th FL
New York, NY 10017

Nicole Gabrielle
Grey Global Group
777 Third Avenue 10th FL
New York, NY 10017

Nina Pratt
Grey Global Group
777 Third Avenue 10th FL
New York, NY 10017

Arista Baltronis
Grey Global Group
777 Third Avenue 10th FL
New York, NY 10017

Asian Talent
Casting Dept
Kang & Lee Advertising Inc
20 Cooper Square
New York, NY 10003

No 8x10s
Deborah Brown
Deborah Brown Casting
160 West End Ave
New York, NY 10023

3 of Us Studios
Donna Grossman
39 W 19th St 12th Fl
New York, NY 10011

Carol Hanzel
Carol Hanzel Casting
48 W 21st St 7th Fl
New York, NY 10011

Print Only
Robin Bobbe
Robin Bobbe Casting
51 W 28th St
New York, NY 10001

Beauty
Grey Global Group
777 Third Avenue 10th FL
New York, NY 10017

Rose Miksis-Zivic
Rotter/Kantor Inc
215 Lexington Ave 14th FL
New York, NY 10016

Abby Casper
Fountainhead Design Inc
33 W 17th St 8th FL
New York, NY 10011

Asian Talent
Casting Dept
House Films
250 W 26th St 3rd FL
New York, NY 10001

VO Only
Sofia Metaxas
Furman Roth Advertising
801 Second Avenue
New York, NY 10017

Kristine Bulakowski
Kristine Bulakowski Casting
Prince St Station, POB 616
New York, NY 10012

Joey Guastella
Joey Guastella Casting
85-10 151st St #5B
New York, NY 11414

Judy Henderson
Judy Henderson & Assoc Casting
330 W 89th St
New York, NY 10024

Ilene Shannon
Shannon's People
27 W 20th St #405
New York, 10011

Tina Sperber
Saatchi & Saatchi Advertising
375 Hudson St
New York, NY 10014

Kirsten Walther
Saatchi & Saatchi Advertising
375 Hudson St
New York, NY 10014

Assoc Creative Director
Bruce Treadwell
Kallier, Phillips, Ross
333 E 38th St
New York, NY 10016

Allen Green
Craig Lechner
Impossible Casting
122 W 26th St 11th FL
New York, NY 10001

Dave Bassiri
Mezzina Brown & Partners
401 Park Ave So 7th FL
New York, NY 10016

Tony Greco
Maria Greco
Maria Greco Casting
630 Ninth Avenue #702
New York, NY 10036

Jimmy Floyd
Jimmy Hank Promotions
209 W 104th St #2H
New York, NY 10025

Barry Shapiro
Herman-Lipson Casting Inc
630 Ninth Ave #1410
New York, NY 10036

Barry Moss
Hughes-Moss Casting Inc
484 W 43rd St #28R
New York, NY 10036

Liz Josefsberg
Inproduction Casting
589 Eighth Avenue 20th FL
New York, NY 10018

Jay Binder
Jack Bowden
Jay Binder Casting
321 W 44th St
New York, NY 10022

Nora Brennan
Nora Brennan Casting
725 West End Avenue Mezz-C
New York, NY 10025

Donald Case
Donald Case Casting Inc
386 Park Avenue So #809
New York, NY 10016

Jeffrey Vigneault
Sharon Chantiles
Chantiles/Vigneault Casting
39 W 19th St 12th Fl
New York, NY 10011

Amy Gossels
Amy Gossels Casting
1382 Third Avenue
New York, NY 10021

Tara Jaffe
Tara Jaffe Casting
235 West 76th Street #2A
New York, NY 10021

Group Effort Films
Jeff Block
Jeff Block Casting
190 Elizabeth St #4R
New York, NY 10012

Toni Roberts
Toni Roberts Casting Inc
150 5th Avenue #309
New York, NY 10011

A Benson
S Dupree/D Kirman
The Casting Connection
199 New Road, Central Square
Linwood, NJ 08221

Kathleen Chopin
Kathleen Chopin Casting
5 W 36th St #602
New York, NY 10018

Laura Richin
Laura Richin Casting
33 Douglas St
Brooklyn, NY 11231

Robyn Todd
Peter Kalin
Kalin/Todd Casting
425 W 58th St #4D
New York, NY 10022

Peter Block
Block Casting
1710 First Avenue Box 170
New York, NY 10128

Joel Carlton
Joel Carlton Casting
212 W 35th St 2nd Fl
New York, NY 10001

Liz Ortiz Mackes
Casting Solutions
231 W 29th St #601
New York, NY 10001

Todd Thaler
Todd Thaler Casting
130 W 57th St #10A
New York, NY 10018

BABIES, CHILDREN & YOUNG ADULTS AGENTS & MANAGERS NY AREA

Claretta King
American Int'l Talent
303 W 42nd St #608
New York, NY 10036

Mary Anne Claro
Mary Anne Claro Agency Inc
1513 Passyunk Avenue
Philadelphia, PA 16145

Bonnie Shumofsky
Abrams Artists & Assoc
275 Seventh Avenue 26th FL
New York, NY 10001

Teens/Y Adults, Lewis Chambers
The Bethel Agency
311 W 43rd St #502
New York, NY 10036

B Halle Feldstein
Cunningham, Escott & Dipene
257 Park Avenue S 9th FL
New York, NY 10010

Young Adults Dept
Agents For The Arts Inc
203 W 23rd St 3rd Fl
New York, NY 10011

B Nancy Carson
Carson-Kolker Agency
250 W 57th St #2030
New York, NY 10107

Ginger Dicce
Ginger Dicce Talent Agency
56 W 45th St #1100
New York, NY 10036

John Richard Sanchez
Allied Artists Mgmt
244 Fifth Avenue E273
New York, NY 10003

B Helene Sokol
Cuzzins Management
250 W 57th St #1001
New York, NY 10107

Young Adult
Acme Talent & Lit. Agency
875 Sixth Avenue #2108
New York, NY 10001

Shirley Faison, Bonnie Deroski
Carson-Alder Agency Inc
250 W 57th St #2030
New York, NY 10107

Herb Rothman
American Talent Mgmt
244 Fifth Avenue #503
New York, NY 10001

Victoria Kress
Don Buchwald Associates
10 E 44th St
New York, NY 10017

Kenneth R Hamilton
Debut Mgmt
2589 Juliet Place Suite 10
Union, NJ 07083

(B) Edie Robb
Creative Mgmt Group
301 W 53rd St #4K
New York, NY 10019

Leslie Zaslower
Cunningham, Escott & Dipene
257 Park Avenue S 9th Fl
New York, NY 10010

Barbara Andreadis
Andreadis Talent Agency
119 W 57th St #711
New York, NY 10019

Sal DiBella
The Actors Club Talent & Mgmt
PO Box 36
Oyster Bay, NY 11771

Teens/Young Adults
Michael Amato Agency
1650 Broadway #307
New York, NY 10019

Tanden R Heyes
Creative Talent Mgmt
91-08 172nd St
Jamaica, NY 11432

Barry Kolker
Carson Organization Ltd
234 W 44th St Penthouse 12
New York, NY 10036

B-Sharon Carry
Carry Company
49 W 46th St
New York, NY 10036

Winter Donaldson
Winter Donaldson Mgmt
33 Cedar Lane
Ossining, NY 10562

Print-Lori Amster
Cunningham, Escott & Dipene
257 Park Avenue S 9th Fl
New York, NY 10010

Ken Nyquist, Dee Mura
Dee-Mura Ent. Inc
269 W Shore Drive
Massapequa, NY 11758

Aggie Gold
Fresh Faces Agency Inc
2911 Carnation Avenue
Baldwin, NY 11510

Paul Donelan
Donelan Management
Suite 15
440 Ninth Avenue 8th Fl
New York, NY 10001

Robert Persad
12 And Up
Electric Talent Inc
172-13 Hillside Avenue #202
Jamaica Estates, NY 11432

Elaine Gordon
Elaine Gordon Management
2942 Harbor Road
Merrick, NY 11566

Patti Fleischer-Print
Generation Model Mgmt
20 W 20th St #1008
New York, NY 10011

Sid Gold
Goldstar Talent Management
850 Seventh Avenue #904
New York, NY 10019

Laurie Kanner
Teens/Young Adults
Kanner Talent Mgmt
3515 Henry Hudson Pkwy Suite
12D

Young Talent-Jana Kogen
Innovative Artists
235 Park Avenue S 7th
New York, NY 10003

Young Talent
Innovative Artists & Ford Models
142 Greene Street
New York, NY 10012

B-TV Dina Torre
Generation TV
20 W 20th St #1008
New York, NY 10011

Randi Ross/Young Adults
EWCR & Associates
311 W 43rd St #304
New York, NY 10036

Arline McGovern
Goodwin/McGovern Mgmt
9 Layton Avenue
Hicksville, NY 11801

B-Robin Dornbaum
Jordan, Gill & Dornbaum Agency
1133 Broadway #623
New York, NY 10010

B-Print, Wendy/Andrea
Ford Children's Division
142 Greene Street
New York, NY 10012

Veronica Goodman
Veronica Goodman Agency
34 North Black Horse Park
Runnemede, NJ 08078

B-Jan Jarrett
Jordan, Gill & Dornbaum
1133 Broadway #623
New York, NY 10010

Children's Dept
McDonald Richards, Inc
156 Fifth Avenue #222
New York, NY 10010

Kim Matuka
Online Talent Group
276 Fifth Avenue #204
New York, NY 10001

B-Children's Dept
Gilla Roos Ltd
16 W 22nd St 3rd Fl
New York, NY 10010

Andree Kaminsky
Fretless Prods. Mgmt
70A Greenwich Avenue #212
New York, NY 10011

B-Shirley Grant
Shirley Grant Mgmt
PO Box 866
Teaneck, NJ 07666

Stanley Kaplan
Stanley Kaplan Talent
139 Fulton St #503
New York, NY 10038

B-Children's Dept
Funnyface Today
381 Park Avenue S
New York, NY 10016

Jano Herbosch
Teens/Y Adults
Herbosh Management
1035 Park Avenue
New York, NY 10028

Lillian Kerr
Lillian Kerr Talent Mgmt
130 Romaine Avenue
Jersey City, NJ 07306

B-18 Yrs, Jeff Mitchell
J Mitchell Management
440 Park Avenue S 11th Fl
New York, NY 10016

B-Dorothy Palmer
Dorothy Palmer
235 W 56th St
New York, NY 10019

Amy R Sussman
Sam's Kids
901 Old Marlton Pike
Marlton, NJ 08053

Donna Gerbino
Innovative Artists
235 Park Avenue S 7th
New York, NY 10003

Saverio Potente/Laurie Potente
Dream vision Ltd
102 Woodbury Road
Hicksville, NY 11803

Legit-Melanie Diaz
Generation TV
20 W 20th St #1008
New York, NY 10011

Barbara Coleman
Innovative Artists
235 Park Avenue S 7th
New York, NY 10003

Jennifer Lambert
Lambert/Gentile Mgmt
22 E 49th St 7th Fl
New York, NY 10017

Eva Cheifetz, Lois Freiman
New York Kids
276 Fifth Avenue 10th Floor
New York, NY 10001

Lori Danziger
Personal Management
40 E 9th Street
New York, NY 10003

Ingrid
Meredith Models East
767 Frederick T
Wyckoff, NJ 07481

B-Young Adults Dept
Fifi Oscard Agency Inc
110 W 40th St #1601
New York, NY 10018

Barclay Tower-Lucille King
Models On The Move
1200 Route 70 #6 POB 4037
Cherry Hill, NJ 08034

Eileen De Nobile
Noble Talent Mgmt
PO Box 66095
Lawrenceville, NJ 08648

Joan Rosenberg
Rosenberg Associates
3 Adams St
Floral Park, NY 11001

Cathy Parker
Cathy Parker Mgmt
PO Box 716
Voorhees, NJ 08043

Richard Realmuto
Richard Realmuto Mgmt
PO Box 760
New York, NY 10033

Michael J Cohen
MJC Entertainment
443 Greenwich St 5th Fl
New York, NY 10013

B-Margaret Martuka
Schuller Talent/NY Kids
276 Fifth Avenue #1001
New York, NY 10001

Children's Dept
The Lyons Group
505 Eighth Avenue #12A
New York, NY 10018

Rosemary P Kassel/Glen West
Padua Kassel Management
210 W 89th St
New York, NY 10024

Children's Dept
Nani/Saperstein Management
16 Penn Plaza Suite 1875
New York, NY 10001

Ayesha Omizelli
Omizelli Management
215 W 101st Street #6A
New York, NY 100125

Latino Actors
Susan Rybin
Rybin Talent Management Ltd
250 W 54th Street #602
New York, NY 10019

Judith A Batista
Parkside Talent Inc
105 Braidburn Road
Florham Park, NJ 07931

Rochelle K Shulman
RKS Management
975 Park Avenue #10C
New York, NY 10028

Suzanne Schachter
Suzelle Enterprises
116 Central Park South Suite 12B
New York, NY 10019

Suzelle Enterprises
100 West 57th Street
New York, NY 10019

DM Kids
Donna Kalakowski
Special Artist Management
555 Shrewsbury Avenue Bldg N
Shrewsbury, NJ 07702

Gail Williams
G Williams Agency
525 South 4th St #365
Philadelphia, PA 19147

Marianne Leone
Terrific Talent Assoc. Inc
419 Park Avenue S #1009
New York, NY 10010

Katherine Atkinson
Washington Square Arts & Film
310 Bowery 2nd FL
New York, NY 10010

Global Network Pictures
244 Madison Avenue
2nd Floor Suite 215
New York, NY 10001

Marilyn Zitner
Marilyn Zitner Mgmt
240 W 35th St
New York, NY 10001

Rick Dorfman
Rick Dorfman Mgmt
270 Lafayette #1308
New York, NY 10001

Tamara Markiowitz
TM Talent
PO Box 72
Marlboro, NJ 07746

Kivi Martin
KMA Agency
11 Broadway #1101
New York, NY 10004

Lynn Tannen
Tannen's Talent LTD
77 Tarrytown Road
White Plains, NY 10607

Lois N Miller
Star Talent Mgmt
682 N Brookside Road #B4
Allentown, P 18106

Teri Bader-Bostaji
Wilhelmina Creative Mgmt
300 Park Avenue S 2nd FL
New York, NY 10010

B-Ellen Gilbert
Abrams Artists & Assoc.
275 Seventh Avenue 26th Fl
New York, NY 10001

Mark Steiner
Steiner Talent Mgmt
1600 Broadway Suite 410
New York, NY 10019

Marlene Wallach
Young Adults
Wilhelmina Creative Mgmt
300 Park Avenue S 2nd Fl
New York, NY 10010

Karen Gunn
Soiree Fair Inc
133 Midland Avenue #10
Montclair, NJ 07042

VOICEOVER AGENTS NY AREA

H Shep Paplin
Agents For The Arts Inc
203 W 23rd St 3rd Fl
New York, NY 10011

Lewis Chambers
Bethel Agency
311 W 43rd St #602
New York, NY 10036

Anita Reilly
Cunningham, Escott & Dipene
257 Park Ave S 9th FL
New York, NY 10010

Oriental Talent-Jadin Wong
Jadin Wong Mgmt
442 W 57th St
New York, NY 10019

Estelle Fusco
Estelle Fusco Mgmt
72 Motiches Road
Lake Grove, NY 11755

Children's Dept
Talentworks Inc
220 E 23rd St #400
New York, 10010

Tobe Gilson
Young Talent Inc
PO Box 792
Hartsdale, NY 10530

Sharon Kutner
S K Talent
PO Box 728
Tenafly, NJ 07670

Jonathan Saul
Abrams Artists
275 Seventh Avenue 26th FL
New York, NY 10001

Barbara Andreadis
Andreadis Talent Agency Inc
119 W 57th St #711
New York, NY 10019

Scott Linder
Don Buchwald Agency
10 E 44th St
New York, NY 10017

Bill Serow
Abrams Artists
275 Seventh Ave 26th FL
New York, NY 10001

Naomi Kolstein
Kolstein Talent Agency
65 Washington Ave. Box 419
Suffern, NY 10901

John Woodward
Nicolosi And Co
150 W 25th St #1200
New York, NY 10001

Jackie Reid
Lil Angels Unlimited
PO Box 312
Tuckahoe, NY 10707

Judith Lesley
Sinclair Management
95 Christopher St #6F
New York, NY 10014

Todd Kijanka-Non Union
Access Talent Inc
37 E 28th St #500
New York, NY 10036

Rachel Sackheim
Atlas Talent
36 W 44th St #1000
New York, NY 10036

Jeff Berber
CTP Casting
207 W 25th St 6th Fl
New York, NY 10001

Pam Lewis-Non Union
Access Talent Inc
37 E 28th St #500
New York, NY 10036

Jaimee Laine
Atlas Talent
36 W 44th St #1000
New York, NY 10036

Wanda King
American Int'l Talent
303 West 42nd St #608
New York, NY 10036

Peter Beilin
Peter Beilin Agency
230 Park Avenue #602
New York, NY 10169

Charles Cowing
Access Talent Inc
37 E 28th St #500
New York, NY 10036

Noel Short
Arcieri & Associates
305 Madison Ave #2315
New York, NY 10036

Mary Anne Claro
Mary Anne Claro Agency
1513 W Passyunk Avenue
Philadelphia, PA 19145

Debra Sherry
Innovative Artists Talent
235 Park Avenue S 7th FL
New York, NY 10003

Stanley Kaplan
Stanley Kaplan Talent Agency
139 Fulton St #503
New York, NY 10038

Ken Franklin
Radioactive Talent Inc
350 Third Avenue Box 400
New York, NY 10010

Allen G. Duncan
Innovative Artists Talent
235 Park Avenue S 7th FL
New York, NY 10003

Steven Arcieri
Arcieri & Associates
305 Madison Ave #2315
New York, NY 10036

Robyn Starr
Don Buchwald Agency
10 E 44th St
New York, NY 10017

Eileen Haves
Acme Talent & Lit. Agency
875 Sixth Ave #2108
New York, NY 10001

Melissa Rollins
Atlas Talent
36 W 44th St #1000
New York, NY 10036

Billy Collura
Cunningham, Escott & Dipene
257 Park Ave S 9th FL
New York, NY 10010

Peter Kalin
Innovative Artists Talent
235 Park Avenue S 7th FL
New York, NY 10003

Oliva Catt
Paradigm Agency
500 Fifth Avenue 37th FL
New York, NY 10110

J Adler
Abrams Artist
275 Seventh Avenue 26th FL
New York, NY 10001

Brooks Barnett
Innovative Artists Talent
235 Park Avenue S 7th FL
New York, NY 10003

Sharon Bierut
Cunningham, Escott & Dipene
257 Park Ave S 9th FL
New York, NY 10010

Michael Amato
Michael Amato Agency
1650 Broadway #307
New York, NY 10019

Nora Brennan
Nora Brennan Casting
752 West End Ave Mezz C
New York, NY 10025

Elsie Stark
Elsie Stark Casting
39 W 19th St 12th FL
New York, NY 10011

Tara Bragine
ICM Inc
40 W 57th St
New York, NY 10019

Mathew Ambrosia-Promos
Paradigm Agency
500 Fifth Avenue 37th FL
New York, NY 10110

Ginger Dicci
Ginger Dicci Talent
56 W 45th St #1100
New York, NY 10036

David Coakley
ICM Inc
40 W 57th St
New York, NY 10019

Ed Batchelor
Paradigm Agency
500 Fifth Avenue 37th FL
New York, NY 10110

Naomi Kolstein
Kolstein Talent Agency
Box 419
65 Washington Avenue
Suffern, NY 10901

Craig Holzberg
Talentworks Inc
220 E 23rd Street #400
New York, NY 10010

Beverly Bishop
Innovative Artists Talent
235 Park Avenue S 7th FL
New York, NY 10003

Marc Guss
William Morris Agency
1325 Sixth Avenue
New York, NY 10019

Tamar Wolbrom-No Tapes
Tamar Wolbrom Inc
130 W 42nd Street #707
New York, NY 10036

Elsie Stark
Elsie Stark Casting
39 W 19th Street 12th Fl
New York, NY 10011

Adam Goodman
Voicehunter.com
10 Sunset Drive
Weston, CT 06883

Steave Garrin-Louise Moore
Videoactive Talent
1780 Broadway #804
New York, NY 10019

Jim Kozloff
VH1/MTV Networks Casting
1633 Broadway 6th FL
New York, NY 10019

Jeb Berstein
Paradigm Agency
500 Fifth Avenue 37th FL
New York, NY 10110

Robert Persad
Electric Talent Inc
172-13 Hillside Avenue #400
Jamaica Estates, NY 11432

Phil Sutfin
ICM Inc
40 W 57th St
New York, NY 10019

Vivkie Barroso-VO/Promos
Paradigm Agency
500 Fifth Avenue 37th FL
New York, NY 10110

Steven Unger-No Tapes
Gage Group Inc
315 W 57th Street #408
New York, NY 10019

Charles Rosen
Charles Rosen Casting
140 W 22nd St 4th Fl
New York, NY 10011

Kathy Wickline
Wickline Casting
1080 N Delaware Ave
Philadelphia, PA 19125

Carole M Duckworth
Dan Duckworth
AAA Voice Casting
123 W 18th St 7th Fl
New York, NY 10011

Maria L Pappalardo
BBDO Inc
1285 Sixth Avenue 6th Fl
New York, NY 10019

Carole Ingber
Ingber & Associates
274 Madison Avenue #1104
New York, NY 10016

David Evans
ICM Inc
40 W 57th St
New York, NY 10019

Dorothy Palmer
Dorothy Palmer Talent
235 W 56th St
New York, NY 10019

VOICEOVER CASTING DIRECTORS NY AREA

Susan Shopmaker
Susan Shopmaker Casting
21 E 26th St 2nd Fl
New York, NY 10036

Jennifer Joy Arnett
Sirius Casting
29 John St PMB 126
New York, NY 10038

Theresa Kiernan
Howard Schartz Recording
420 Lexington Avenue #1934
New York, NY 10170

Breanna Benjamin
Breanna Benjamin Casting
PO Box 36-20967
New York, NY 10129

Carol Hanzel
Carol Hanzel Casting
48 W 21st St 7th FL
New York, NY 10022

Liz Ortiz Mackes
Casting Solutions
231 W 29th Street #601
New York, NY 10001

Judy Keller-No 8x10's
Judy Keller Casting
140 W 22nd St 4th FL
New York, NY 10011

SM Gish/Susan Gish
Philadelphia Casting Company
114 Chestnut Street #rd Fl
Philadelphia, PA 19106

Donna De Seta
Donna De Seta Casting
525 Broadway 3rd FL
New York, NY 10012

3 of Us Studios-Joan Lynn
Joan Lynn Casting
39 W 19th St 12 FL
New York, NY 10011

Tom Harris
Donald Case Casting Inc
386 Park Avenue S #809
New York, NY 10016

Liz Hosefsberg
Inproduction Casting
580 Eighth Avenue 20th Fl
New York, NY 10018

Joanne Pasciuto-No 8x10s
Joanne Pasciuto Casting
17-08 150th St
Whitestone, NY 11357

Tricia Lentini
Avrett, Free & Ginsberg
800 Third Avenue 35th Fl
New York, NY 10022

Merry Delmonte
Merry L Delmonte Casting
575 Madison Avenue 10th FL
New York, NY 10022

Staacy L Seidel
Liz Lewis Casting
129 W 20th Street
New York, NY 10011

Don Case
Donald Case Casting, Inc
386 Park Avenue S #809
New York, NY 10016

Diane Heery
Heery Casting
230 North 2nd St #1A
Philadelphia, PA 19106

Matthew Messinger-No 8x10s
Matthew Messinger Casting
244 W 72nd St
New York, NY 10023

Sue Crystal
Sue Crystal Casting
251 W 87th Street #26
New York, NY 10024

Rob Holt
Mike Lemon Casting
413 N 7th Street #602
Philadelphia, PA 19123

Toni Roberts
Toni Roberts Casting Inc
1133 Broadway #630
New York, NY 10010

Ana Melara
Wing Latino Group
114 Fifth Avenue 11th FL
New York, NY 10003

Lesley Freedman/McHale Barone
30 Irving Place 6th Fl
New York, NY 10003

Sharon Chantiles
Chantiles/Vigneault Casting
39 W 19th St 12th Fl
New York, NY 10011

Casting Dept
Kipperman Casting Inc
12 W 37th St 3rd Fl
New York, NY 10018

Jennifer Sukup
Pomann Sound
2 W 46th St Ph
New York, NY 10036

Casting Dept
Godlove Casting
151 W 25th St 11th Fl
New York, NY 10001

Jeff Barber/Angela Montalbano
MBC Casting
325 W 38th Street
New York, NY 10018

AD AGENCIES NEW YORK PRODUCTION COMPANIES

Sofia Metaxas
Furman Roth Advertising
801 Second Avenue 15th Fl
New York, NY 10017

John King
John King Productions
244 Madison Avenue #393
New York, NY 10016

Tina Sperber
Saatchi & Saatchi Advertising Inc
375 Hudson Street
New York, NY 10014

Jim Kennelly-VO
John Lotas Productions
1123 Broadway #1218
New York, NY 10010

Nina Pratt
Grey Global Group
777 Third Avenue
New York, NY 10017

Teresa Wolf
Schiowitz/Clay/Rose
165 W 46th Street #1210
New York, NY 10036

Margaret Matuka
Schuller Talent
276 Fifth Avenue #1001
New York, NY 10001

Rachel Sheedy
Don Buchwald associates
10 E 44th Street
New York, NY 10017

Gail Williams
Gail Williams Agency
525 S 4th Street #365
Philadelphia, PA 19147

Marvin A Josephson
Gilla Roos Ltd
16 W 22nd Street 3rd Fl
New York, NY 10010

Honey Raider
Talent Reps
20 E 53rd Street #2A
New York, NY 10022

Nancy Espinal-VO Only
TBWA/Chiat/Day
488 Madison Avenue
New York, NY 10022

Lew Sherwood
Warren, Kremer & Paino
2 Park Avenue #1400
New York, NY 10016

NEW YORK (AEA) EQUITY AGENTS

Jay Kane
Talentworks Inc
220 E 23rd Street #400
New York, NY 10010

Sheldon Lubliner
Professional Artists
321 W 44th Street #605
New York, NY 10010

Bill Iannone
Talent Network Group
111 E 22nd Street
New York, NY 10010

Roberta Warren/Babs Zimmerman
Babs Zimmerman Productions
305 E 86th Street 17 FW
New York, NY 10028

Musical Theatre
Christianne Weiss
International Creative Mgmt
40 w 57th Street
New York, NY 10019

Joanna Kaplowitz
Peter Strain & Associates
1501 Broadway #2900
New York, NY 10036

Dan Aron
No Soap Productions
630 Ninth Avenue #602
New York, NY 10036

Dave Bennett/Bruce Dean
Talent House Agency
311 43rd Street #602
New York, NY 10036

William Chaver
Writers & Artists Agency
19 W 44th Street #1000
New York, NY 10036

Boaty Wright
International Creative Mgmt
40 W 57th Street
New York, NY 10019

Peter Strain
Peter Strain & Associates
1501 Broadway #2900
New York, NY 10036

Ann Steele
Ann Steele Agency
331 W 42ndf Street 18th Fl
New York, NY 10036

Empire State Bldg/Sharon Carry
Carry Company
350 Fifth Avenue #1702
New York, NY 10118

Ann Wright
Ann Wright Representatives
165 W 46th Street #1105
New York, NY 10036

Kevin Thompson
Schiowitz/Clay/Rose
165 W 46th Street #1210
New York, NY 10036

Alan Willig
Don Buchwald Associates
10 E 44th Street
New York, NY 10017

Helen T. Shabason
International Creative Mgmt
40 W 57th Street
New York, NY 10019

Archer King
Archer King Ltd
317 W 46th Street 3A
New York, NY 10036

Lionel Larner
Lionel Larner Ltd
119 W 57th Street #1412
New York, NY 10019

Meg Pantera
Meg Pantera Agency
1501 Broadway Suite 1508
New York, NY 10036

Dianne Busch
Leading Artists Inc
145 W 45th Street #1204
New York, NY 10036

Scott Lonker
William Morris Agency
1325 Sixth Avenue
New York, NY 10019

Judy Boals
Judy Boals Inc
208 W 30th Street #401
New York, NY 10001

Bernard Liebhaber
Bernard Liebhaber Agency
352 Seventh Avenue
New York, NY 10036

Tim Stone
Stone-Manners Agency
900 Broadway #803
New York, NY 10003

Steve Carson
Carson-Kolker Organization
W 44th Street #902
New York, NY 10036

Dale Lally
Lally Talent Agency
630 Ninth Avenue #800
New York, NY 10036

Biff Liff
William Morris Agency
1325 Sixth Avenue
New York, NY 10019

Ellison K Goldberg
Kerin-Goldberg & Associates
155 E 55th Street #5D
New York, NY 10022

Penny Luedtke
The Luedtke Agency
1674 Broadway #7A
New York, NY 10019

Tom Ingregno
Omnipop Inc
55 W Old Country Road
Hicksville, NY 11801

Ronald Ross
Kerin-Goldberg & Associates
155 E 55th Street #5D
New York, NY 10022

Joel Carlton
The Ludtke Agency
1674 Broadway #7A
New York, NY 10019

Lori Swift
Kazarian-Spencer & Associates
162 W 56th Street 3rd Fl
New York, NY 10019

Bruce Levy
Bruce Levy Agency
311 W 43rd Street #602
New York, NY 10036

Toni Sipka
Nouvelle Talent Inc
20 Bethune Street #5A
New York, NY 10014

Naomi Kolstein
The Kolstein Talent Agency
65 Washington Avenue
Suffern, NY 10901

Mark Subias
William Morris Agency
1325 Sixth Avenue
New York, NY 10019

Richard Schmenner
Paradigm Agency
500 Fifth Avenue 37th FL
New York, NY 10110

B Lynne Jebens
The Krasny Office
1501 Broadway #1510
New York, NY 10036

David Kalodner
William Morris Agency
1325 Sixth Avenue
New York, NY 10019

Rosanne Quezada
Paradigm Agency
500 Fifth Avenue 37th FL
New York, NY 10110

Jean Nicolosi
Nicolosi and Company
150 W 25th Street #1200
New York, NY 10001

Mary Anne Claro
Mary Anne Claro Talent Agency Inc
1513 Passyunk Avenue
Philadelphia, PA 19145

Randi Ross
EWCR & Associates
311 W 43rd Street #304
New York, NY 10036

Mary Harden
Harden-Curtis Associates
850 Seventh Avenue #903
New York, NY 10019

Corey Smith
DDO Artists Agency
116 W Houston Street 2nd Fl
New York, NY 10012

William Butler
Gersh Literary & Talent Agency
41 Madison Avenue 33rd Fl
New York, NY 10010

George Lutsch
Henderson/Hogan Agency
850 Seventh Avenue #1003
New York, NY 10019

Bonnie Bernstein
Endeavor Talent Agency
152 W 57th Street
New York, NY 10019

Peggy Hadley
Peggy Hadley Enterprises Ltd
250 W 57th Street #2317
New York, NY 10107

Francis Delduca
Fifi Oscard Agency Inc
110 W 40th Street #1601
New York, NY 10018

Steven Stone
Cornerstone Talent Agency
37 W 20th Street #1108
New York, NY 10011

Rhonda Price
Gersh Literary & Talent Agency
41 Madison Avenue 33rd Fl
New York, NY 10010

Jerry Hogan
Henderson/Hogan Agency
850 Seventh Avenue #1003
New York, NY 10019

Margaret Emory
Dulcina Eisen Associates
154 E 61st Street
New York, NY 10021

Veronica Goodman
Veronica Goodman Agency
34 North Black Horse Park
Runnemede, NJ 08078

Deborah Coleman
Coleman-Rosenberg
155 E 55th Street #5D
New York, NY 10022

Karen S Garber
The Garber Agency
2 Penn Plaza #1910
New York, NY 10121

Paul Hilepo
Hartig-Hilepo Agency Ltd
156 Fifth Avenue #1018
New York, NY 10010

Michelle Gerard
Douglas, Gorman, Rothacker & Wilhelm
1501 Broadway
New York, NY 10036

Jason Gutman
Gersh Literary & Talent Agency
41 Madison Avenue 33rd Fl
New York, NY 10010

Christopher Freer
Clear Talent Group
440 Ninth Avenue 8th Fl
New York, NY 10001

Philip Adelman
The Gage Group
315 W 57th Street #408
New York, NY 10019

Michael Kirsten
Kirsten Manages Literary Agency
156 Fifth Avenue #1018
New York, NY 10010

Flo Rothacker
Douglas, Gorman, Rothacker & Wilhelm
1501 Broadway
New York, NY 10036

Jennifer Konawal
Gersh Literary & Talent Agency
41 Madison Avenue 33rd Fl
New York, NY 10010

Jack Menashe
Independent Artists Agency
159 W 25th Street #1011
New York, NY 10001

Carole Davis
Eastern Talent Alliance Inc
1501 Broadway #404
New York, NY 10036

Robert Attermann
Abrams Artists
275 Seventh Avenue 26th Fl
New York, NY 10001

Barbara Andreadis
Andreadis Talent Agency Inc
119 W 5th Street #711
New York, NY 10019

Robert Freedman
Robert A. Freedman Agency
1501 Broadway #2301
New York, NY 10036

Margi Rountree
Bret Adams Inc
448 W 44th Street
New York, NY 10036

David A Grunes
Artists Group East
1650 Broadway #711
New York, NY 10019

David Krasaner/Musical Theatre
Bloc Agency
41 E 11th Street 11th Fl
New York, NY 10003

Ken Melamed
Bret Adams Inc
448 W 44th Street
New York, NY 10036

Richard Astor
The Richard Astor Agency
250 W 57th Street #2014
New York, NY 10107

Tony Cloer
Blue Ridge Entertainment
41 Union Square West #809
New York, NY 10003

Nina Shreiber
ACME NY
845 Sixth Avenue #2108
New York, NY 10001

Cynthia Katz
Artists Group East
1650 Broadway #610
New York, NY 10019

Lewis Chamberg
The Bethel Agency
311 W 43rd Street #602
New York, NY 10036

Carole J Russo
Agents For The Arts
203 W 23rd Street 3rd Fl
New York, NY 10011

Mari Cronin
Ann Elmo Agency
60 E 42nd Street
New York, NY 10165

Joanne Nici
Don Buchwald Associates
10 E 44th Street
New York, NY 10017

Claretta King
American Intl Talent
303 W 42nd Street #608
New York, NY 10036

David Krinsky
Earth Tracks Artists Agency
4809 Avenue N #286
Brooklyn, NY 11234

NEW YORK AREA PRINT AGENCIES MD, VA, PA, MA, NJ, CT

John Camillieri
Barry Haft Brown Artists
165 W 45th Street #908
New York, NY 10036

Harry Abrams
Abrams Artists
275 Seventh Avenue 26th Fl
New York, NY 10001

Beverly Anderson
1501 Broadway #2008
New York, NY 10036

Mark Redanty
Bauman, Redanty & Shaul
250 W 57th St.
New York, NY 10107

Jill McGrath
Abrams Artists
275 Seventh Avenue 26th Fl
New York, NY 10001

Jennifer Lamaj
Beverly Anderson
1501 Broadway #2008
New York, NY 10036

Charles Bodner
Bauman, Redanty & Shaul
250 W 57th St.
New York, NY 10107

Mary Guzowski/Genine Esposito
Abrams Artists
275 Seventh Avenue 26th Fl
New York, NY 10001

Stephanie Bellarosa
Cunningham, Escott & Dipene
257 Park Avenue S 9th Fl
New York, NY 10010

Carmen Marufo/Parts Division
Arcieri & Associates
305 Madison Avenue #2315
New York, NY 101065

Robert Persad
Electric Talent Inc
172-13 Hillside Avenue #202
Jamaica Estates, NY 11432

K Stuart/N Annan/Charlie Winfield
Funnyface Today Models
381 Park Avenue S 8th Fl
New York, NY 10016

Lynne Carole
Barbizon Agency
16 Penn Plaza Suite 1205
New York, NY 10001

Women Only, Teen/Young Adults
Elite Model Management (F)
111 E 22nd Street
New York, NY 10010

Ron Sklar
Fusion Models & Management
95 Morton Street
New York, NY 10014

Scan Powers/Rita Powers
295 Greenwich Street #513
New York, NY 10007

Epitome/Mike
E&M Agency
208-76 Grand Central Pkwy #3B
Queen Village, NY 11427

Kirsten Weaver
ID Model Mgmt (F)
110 Greene Street Suite 702
New York, NY 10012

Teens/Young Adults
Boss Models (F)
1 Ganevoort Street
New York, NY 10014

Justin Brown
5A Models Management
1560 Broadway #1110
New York, NY 10036

Teens/Young
IMG Models (F)
304 Park Avenue Sout 12th FL
Penthouse North
New York, NY 10010

231 Real People
Jenevieve Brewer
Carson-Kolker Agency
W 44th Street #902
New York, NY 10036

Gene Roseman
Gene Flaunt Agency
114 E 32nd Street
New York, NY 10016

Empire State
William Michael Carpenter
Grace Del Marco Model & Talent
350 Fifth Avenue #3110
New York, NY 10118

New Faces
Click Models & Management
129 W 27th Street Penthouse
New York, NY 10001

Ford Models (F)
142 Greene Street 4th Fl
New York, NY 10012

Lauren Green
Lauren Green Model Mgmt
156 Fifth Avenue #822
New York, NY 10038

Ray Volant/Commercial Print
Ford Models
142 Greene Street 4th Fl
New York, NY 10012

New Talent
Adele's Kids & Adults
33 Rupert Avenue
Staten Island, NY 10314

Oscar Garnica
Cunningham, Escott & Dipene
257 Park Avenue S 9th Fl
New York, NY 10010

Susan Goldfarb/Commercial Print
Ford Models
142 Greene Street 4th Fl
New York, NY 10012

Mike Amato
Mike Amato Agency
1650 Broadway #307
New York, NY 10019

Bruce Baruch Kivo
Eastwood Talent Group
208 W 30th Street #202
New York, NY 10001

Heather Finn/Teens/Young Adults
Frontier Booking Intl Inc
1560 Broadway #1110
New York, NY 10036

Teens/Young Adults
Karin Models (F)
6 W 14th Street 3rd FL
New York, NY 10013

Karen Apicella
Model Services Inc
570 Seventh Avenue #702
New York, NY 10018

Kivi Martin
KMA Agency
11 Broadway #1101
New York, NY 10004

New Faces
Lyons Group Inc
505 Eighth Avenue 12A Fl
New York, NY 10018

Elise Brown
Renny Model Mgmt
352 Seventh Avenue
New York, NY 10010

Dorothy Palmer
Dorothy Palmer Agency
56th Street
New York, NY 10019

Sharon Reich/Rene Gonzalez
Reich New York
23 W 26th Street GR FL
New York, NY 10010

John King
John King Productions
244 Madison Avenue #393
New York, NY 10016

Print Work
R & L Model Mgmt
203 W 23rd Street #400
New York, NY 10011

Kimberly Thompson
Thompson Model & Talent (F)
50 W 34th Street #6C-6
New York, NY 10001

Teens/Young Adults
Next Model Management (F)
23 Watts Street
New York, NY 10013

Beauty/Fashion
Jacqueline Love
25 Post Street #2W
Yonkers, NY 10705

Gary Bertalovitz
McDonald/Richards Inc
232 Madison Avenue
New York, NY 10016

Samanth Roos/David Roos
Gilla Roos Ltd
16 W 22nd St 3rd Fl
New York, NY 10010

Print Work
Parts Models
PO Box 7529 Flr Sta
New York, NY 10150

Rachael Mitchell
Saatchi Management
13 Crosby Street #301
New York, NY 10013

M Gingold/P Carmody
Life Styles Talent Management
900 Broadway #805
New York, NY 10003

Anne Webb
Talent Express
PO Box 280522
Queens Village, NY 22528

Shawn King-All Types
King Studios
467 E 157th Street
Bronx, NY 10451

Teens/Young Adults
Pauline Bernatchez
Pauline's Talent Company (F)
379 W Broadway #502
New York, NY 10012

Rick Miller
Rick Miller Agency Inc
22 W 21st Street #403
New York, NY 10010

Print Work
Fifi Oscard Agency Inc
110 W 40th Street #1601
New York, NY 10018

Print Work
Talentworks Inc
220 E 23rd Street #400
New York, NY 10010

Scott Powers
Scott Powers Productions
22 W 21st Street #405
New York, NY 10010

Ramona Pitera
Gilla Roos Ltd
16 W 22nd St 3rd Fl
New York, NY 10010

Cherie Pourtabib/Women Only
Wilhelmina Models
300 Park Avenue S 2nd Fl
New York, NY 10013

Jennifer Katz/Men Only
Wilhelmina Models
300 Park Avenue S 2nd Fl
New York, NY 10013

Ann Wright
Ann Wright Reps Inc
165 W 46th Street #1105
New York, NY 10036

Patricia Pinto
La Crème Centre For Modeling Inc
PO Box 131
Union, NJ 07083

Esther Johnston
Johnston Agency
50 Washington Street
South Norwalk, CT 06854

New Talent
Christine Models
43 Rock Road
Wayne, NJ 07470

Heather Wright
Pro Model & Casting Agency
1244 Hamilton Street
Philadelphia, PA 18102

T Denard/S Garrett
McDonald/Richards Inc
5 River Road #317
Wilton, CT 06897

Print Work
Veronica Goodman
34 N Black Horse Park
Runnemede, NJ 08078

Ellen Sciblo
Reinhard Inc
2021 Arch Street 4th Fl
Philadelphia, PA 19103

David McDermott/Adult Division
Jordan, Gill/Dornbaum Agency
1133 Broadway #623
New York, NY 10010

Barclay Towers/Lucille King
Models On The Move
1200 Route 70 #6 POB 4037
Cherry Hill, NJ 08034

Linda Wisch
Plaza 7
160 N Gulph Road
King of Prussia, PA 19406

Print Work
Jennifer Models
755 Montauk Highway #2
Oakdale, NY 11769

Kristen Paladino
Paladino Casting
PO Box 475
Leonia, NJ 07605

Print Work
Bowman Agency
PO Box 4071
Lancaster, PA 17601

Joan Gilbride/Elaine Gordon
Elaine Gordon Model Mgmt
2942 Harbor road
Merrick, NY 11566

Mary Anne Claro
The Claro Agency
1513 W Passyunk Avenue
Philadelphia, PA 19145

Terra In The Spotlight
2608 Columbia Avenue
Lancaster, PA 17603

Pam Young
Allison Daniels Models & Talent
1260 New Britain Avenue
W Hartford, CT 06110

Wm McCullough
McCullough Models Inc
8 S Hanover Avenue
Margate, NJ 08402

Stephen Black
The Talent Group
2820 Smallman Street
Pittsburgh, PA 15222

Linda Gagliardi/Greer Lange
Greer Lange Talent Agency
3 Bala Plaza West #201
Balas Cynwyd, PA 19004

Model Club Inc
329 Columbus Avenue
Boston, MA 02116

Ingrid Schmerling
Meredith Models East
767 Frederick Ct
Wyckoff, NJ 07481

Pam Lankford
Expressions Model & Talent
220 Church Street
Philadelphia, PA 19106

Marcia Weissman
Marcia Weissman Model Mgmt
80 Varick Street #9A
New York, NY 10013

S Norell/D Brown
Axis Models
46 Church Street
Montclair, NJ 07042

Nancy Mac Mahon
Frame Mgmt
12 W Willow Grove Avenue
PMB 189
Philadelphia, PA 19118

Debra Docherty
Docherty Inc
109 Market Street
Pittsburgh, PA 15222

Scott Goodhue/Martha Royal
Taylor Royal Agency
6247 Falls Road
Baltimore, MD 21209

D Wittmer/C Ness
Central Casting/No Mail
Washington, DC 20003
202-310-4700

Darlene/TV
The Saatchi Agency
9868 Main Street
Fairfax, VA 22031

Kathy Baxter
The Models Group
381 Congress Street 5th Fl
Boston, MA 02110

Print Work
Images Mgmt & Talent Agency
30 E 20th Street 6th Fl
New York, NY 10003

Kristine
Central Casting/No Mail
Washington, DC 20003
202-310-4700

Maggie Trichon
Maggie Inc
35 Newbury Street
Boston, MA 02116

Print Work
Tannen's Talent Ltd
77 Tarrytown Road
White Plains, NY 10607

Jason
The Saatchi Agency
9868 Main Street
Fairfax, VA 22031

Brain Bullock/Patricia Bullock
The Bullock Agency Inc
5200 Baltimore Avenue #102
Hyattsville, MD 20781

Made in United States
Cleveland, OH
06 November 2026

47787695R00127